DPH MATHEMATICS SERIES

TEXT BOOK OF TRIGONOMETRY

By

A.K. Sharma

DISCOVERY PUBLISHING HOUSE
NEW DELHI-110002

First Published – 2004
Reprinted – 2025

ISBN: 978-81-7141-904-3

Text Book of Trigonometry

Published by:
DISCOVERY PUBLISHING HOUSE
4383/4B, Ansari Road, Darya Ganj
New Delhi-110 002 (India)
Phone: +91-11-23279245; 23253475; 43596065
Mobile: +91 9811179893 / +91 9871656464
E-mail: discoverybooksindia@gmail.com
orderdphbooks@gmail.com
namitwasan9@gmail.com
web: www.discoverypublishinggroup.com

Printed at:
Infinity Imaging Systems
Delhi

Preface

This book 'Text book of Trigonometry' has been specially written to meet the requirement of Degree and Honours students of various Indian Universities. The subject matter of this book has been discussed in such a simple way that the students find no difficulty to understand. Each chapter of this book contains complete theory and large number of solved example.

We hope this book will be found useful by the students and teachers on the various Indian Universities. We will appreciate any suggestion for the improvement of the book.

A.K. Sharma

Contents

Pages

Preface

1. **Complex Numbers** 1

Complex Numbers, Set of Complex Numbers, Equality of Two Numbers, Negative of a Complex Number, Zero Complex Number, Conjugate of a Complex Number, Properties of Conjugate, Reciprocal of a Complex Number, Logarithm of Complex Number, Method to Find $(X + IY)A^{+IB}$, Modulus and Argument, Polar Representation of a Complex Number, Four Fundamental Operations, Important Results, De-Moivre's Theorem, Some Results Based on De-Moivre's Theorem, Cube Roots of Unity, Properties of Cube Roots of Unity, Important Relations in Complex Number, Properties, De-Moivre's Theorem.

2. **Exponential, Trigonometric and Hyperbolic Functions of a Complex Variable (Separation into Real and Imaginary Parts)** 71

The Exponential Functions of a Complex Variable, Index Law for the Exponential Functions, Trigonometrical Functions or Circular Functions of a Complex Variable, Euler's Theorem, Periodicity of Functions, De Moivre's Theorem for Complex Argument, Some Standard Trigonometrical Results for Complex Arguments, Hyperbolic Functions, Relations Between Hyperbolic and Circular Functions, Properties of Hyperbolic Functions, Expansions in Series for sinh x and cosh x, Periods of Hyperbolic Functions, Separation into Real and Imaginary Parts.

3. **Logarithms of Complex Numbers** 101

Logarithms in the Set of Real Numbers, Logarithms of Complex Numbers, Principal and General Values of Logarithm of a Non-Zero Complex Number, Properties of the Logarithmic Function, Working Rule to Evaluate Log (x + iy) *i.e.*, To Express Log (x + iy) in the Form A + iB, Logarithm of a Positive Real Number in the Set of Complex Numbers, Logarithm of a Negative Real Number, The General Exponential Function a^z, To Separate $(\alpha + i\beta)^{p + iq}$ into Real and Imaginary Parts.

4. Inverse Circular and Hyperbolic Functions of Complex Numbers 123

Inverse Circular Functions, General and Principal Values of Inverse Circular Functions, Relations Between Inverse Functions, Some Important Results about Inverse Functions, Inverse Circular Functions of Complex Numbers, Inverse Hyperbolic Functions, Relations Between Inverse Hyperbolic Functions and Inverse Circular Functions.

5. Expansion of Some Trigonometrical Functions 141

Introduction, Expansion of $Cos^N \Theta$ in Series of Cosines of Multiple of Θ, n being a Positive Integer, Expansion of $Sin^N \Theta$ in a Series of Cosines and Sines of Multiples of Θ, According as n (A Positive Integer) is Even or Odd, Expansion of $Cos^N \Theta$ and $Sin^N \Theta$ in Terms of Sines and Cosines of Multiple of Θ, N being a Positive Integer, Expression for Tan $(\Theta_1 + \Theta_2 ... \Theta_N)$ in Terms of Tan Θ_1; Tan Θ_2 ..., Tan Θ_N,.

6. Geogory's Series and Trigonometrical Expansions 165

Geogory's Series, General Theorem of Geogory's Series, Value of p, Expansions.

7. Summation of Trigonometrical Series 193

Introduction, General C + is Method, Use of Geometric Series, Summation of Series,

1

Complex Numbers

1.1 COMPLEX NUMBERS

A complex number is a number which can be written in the form a + ib, where a and b are real numbers and i = $\sqrt{-1}$. It is generally denoted by single letter z. The real number a is called the real part of the complex number z = a + ib and the real number b is called the imaginary part of z.

A complex number is also defined as ordered pair (a, b) of real numbers. Thus, if z if a complex number, then

$$z = a + ib = (a, b) \text{ for some } a, b \in R.$$

we can write $z = \text{Re}(z) + i\,\text{Im}(z)$

1.2 SET OF COMPLEX NUMBERS

The set of complex numbers is denoted by C, where

$$C = \{z \mid z = a + ib, a, b, \in, R\}$$

or $\quad C = \{z \mid z = (a, b), ab \in R\}$

In other words the Cartesian product R × R Consisting of the ordered pairs of real numbers is called the set of complex numbers

1.3 EQUALITY OF TWO NUMBERS

Two complex numbers are said to be equal if and only if their real and imaginary parts are separately equal, *i.e.*,

$$a + ib = c + id$$

$$\Leftrightarrow \quad a = c \text{ and } b = d.$$

1.4 NEGATIVE OF A COMPLEX NUMBER

The complex $-z = -a - ib$ is called the negative of the complex number $z = a + ib$ and vice–versa.

1.5 ZERO COMPLEX NUMBER

The complex number $z = a + ib$, a, b, $\in$ R is said to be zero complex number or zero of C if only a = O and b = O.

1.6 CONJUGATE OF A COMPLEX NUMBER

Two complex numbers (a, b) and (a, – b) are said to be complex conjugate to each other. In other words, the two complex numbers of the type a + ib and a — ib are said to be complex conjugate to each other. The conjugate of a number $z = a + ib$ is denoted by $z = a - ib = \text{Re}(z) - i\,\text{Im}(z)$.

1.7 PROPERTIES OF CONJUGATE

1. $\overline{(\overline{z})} = z$
2. $z + \overline{z} = (a + ib) + (a - ib) = 2a = 2$ Real z or 2 Re z (where real z means the real part of z)
3. $z - \overline{z} = (a + ib) - (a - ib) = 2ib = 2i \times$ Imaginary z or $-2i$ Im z

   (Where imaginary z or Im z Means the imaginary part of z)
4. $z..\overline{z} = (a + ib)(a - ib) = a^2 + b^2 = \overline{z}.z$

   It may be noted that the conjugate of a real number is the real number it self.

1.8 RECIPROCAL OF A COMPLEX MUMBER

Let $z = a + ib$ be a non-zero complex number. Then $\frac{1}{z} = \frac{1}{a+ib}$

$$\Rightarrow \frac{1}{a+ib} \times \frac{a-ib}{a-ib}$$

$$\Rightarrow \frac{a-ib}{a^2 - i^2 - b^2} = \frac{a-ib}{a^2+b^2} = \frac{a}{a^2+b^2} + \frac{i(-b)}{a^2+b^2}$$

Clearly, 1/2 is equal to the multiplicative inuerse of z. Also

$$\frac{1}{z} = \frac{a-ib}{a^2+b^2} + i\frac{(-\text{Im}(z))}{|z|^2} = \frac{\overline{z}}{|z|^1}$$

Thus, the multiplicative inverse of a non-zero complex number z is same as its reciprocal and is given by

$$\frac{\text{Re}(z)}{|z|^2} + i\frac{(-\text{Im}(z))}{|z|^2} = \frac{\bar{z}}{|z|^2}$$

1.9 LOGARITHM OF COMPLEX NUMBER

In order to find log (x + iy) we write log(x + iy) = a + ib

$x + iy = e^{a+ib} = e^a [\cos b + i \sin b]$

$= e^a [\cos (2k\pi + b) + i \sin (2k\pi + b]$

$\therefore\ e^x \cos (2k\pi + b) = x$ and $e^a \sin (2k\pi + b) = y$

Solve for a and b $\therefore\ e^{2a} = x^2 + y^2$

or $a = 1/2 \text{ In } (x^2 + y^2)$, $\tan (2k\pi + b) = (y/x)$

When k = 0 corresponding values of a and b are refered to as principale values.

1.10 METHOD TO FIND $(X + IY)^{A+IB}$

To evaluate $(x + iy)^{a+ib}$ we write

$c + id = (x + iy)^{a+ib}$

$\log (c + id) = (a + ib) \log (x + iy)$

Now evaluate log (x + iy) and then solve

$c + id = e^{(a+ib) \log (x+iy)}$.

1.11 MODULUS AND ARGUMENT.

The moduls of a complex number z = a + ib is denoted by | z | and is defined as

$$|z| = \sqrt{a^2 + b^2} = \sqrt{\{\text{Re}(z)\}^2 + \{\text{Im}(z)\}^2}$$

Clearly $|z| \geq 0$ for all $z \in c$.

Argument

If $r \neq O$, the angle θ such that $\theta = \tan^{-1}(y/x)$ is called the argument or amplitude of the complex number z = x + iy and is denoted by θ = arg. z.

It is important to note that if θ is an argument of z, then $2\pi + \theta$ os also an argument of z as $\cos (2\pi + \theta) = \cos \theta$, $\sin (2\pi + \theta) = \sin \theta$.

In general $2n\pi + \theta$, $n \in N$, is also an argument of z, if θ is it argument. The value of θ which lies between π and $-\pi$ is called the principal value of the argument.

Unimodular Complex Number

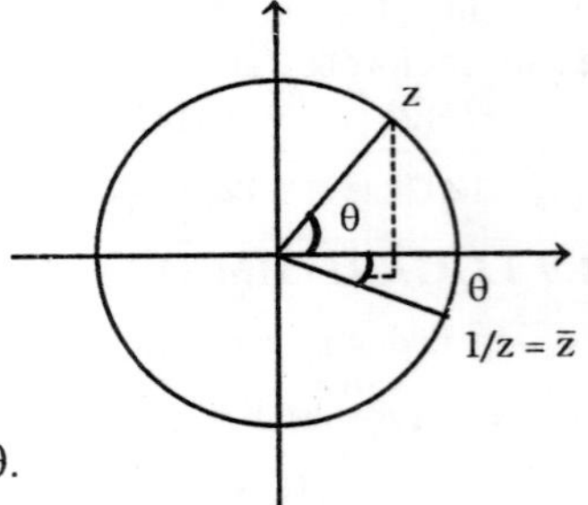

A complex number z such that $| z | = 1$ is soid to be unimodular complex number since $| z | = 1$ z lies on a circle of radius 1 unit and centre (0, 0) If

$$z = 1 \Rightarrow z = \cos\theta + i \sin\theta$$

$$\Rightarrow 1/z = (\cos\theta + \sin\theta)^{-1} = \cos\theta - i \sin\theta.$$

1.12 POLAR REPRESENTATION OF A COMPLEX NUMBER

This representation $z = r(\cos\theta + i\sin\theta)$, where $r = \sqrt{x^2+y^2}$ and $\theta = \tan^{-1}(y/x)$ is known as the trigonometrical or Polar representation of a complex number z. The non- negative real number $\sqrt{x^2+y^2}$ called the modulus or absolute value of complex number $z = a + ib$ and is denoted by $| z |$ *i.e.*, $| z| = \sqrt{x^2+y^2}$ Also $z = x^2 + y^2 = | z | z$

1.13 FOUR FUNDAMENTAL OPERATIONS

If $z_1 = a + ib_2$ and $z_2 = c + id$, then

$z_1 + z_2 = (a + c) + (b + d)i$ (Addition)

$z_1 - z_2 = (a - c) + (b - d)i$ (Subtraction)

$z_1 z_2 = (ac - bd) + (ad + bc)i$ (Multiplication)

$$\frac{z_1}{z_2} = \frac{a+bi}{c+di} = \frac{(a+bi)(c+di)}{c^2+d^2} \quad \text{(Division)}$$

$$= \left[\frac{ac+bd}{c^2+d^2}\right] + \left(\frac{bc=ad}{c^2+c^2}\right)i$$

1.14 IMPORTANT RESULTS

1. $| z | = | - z | = | \bar{z} | = | -\bar{z} |$.
2. $z + \bar{z} = 2 \operatorname{Re}(z)$.
3. $z - \bar{z} = 2i \operatorname{Im}(z)$.
4. $z. \bar{z} = | z | 2$.
5. $| z_1 + z_2 | \leq | z_1 | + z_2 |$.

6. $| z_1 - z_2 | \geq | z_1 | - z_2 |$.
7. $| z_1 . z_2 | = | z_1 | . z_2 |$.
8. $\left|\frac{z_1}{z_2}\right| = (1z_1 |/ 1z_2|). (z_2 \neq 0)$.
9. $\text{Arg}\ (z_1 . z_2) = \text{Arg}\ (z_1) + \text{Arg}\ (z_2)$
10. $\text{Arg}\left[\frac{z_1}{z_2}\right] = \text{Arg}\ (z_1) — \text{Arg}\ (z_2)\ (z_2 \neq 0)$.

1.15 DE-MOIVRE'S THEOREM

Statement : If n is an integer, positive or negative, or a rational, number, then the value or one of the value or one or the values of $(\cos\theta + i \sin\theta)"$ is $(\cos n\theta + i \sin n\theta)$.

1.16 SOME RESULTS BASED ON DE-MOIVRE'S THEOREM

1. $(\cos\theta + i \sin\theta)" = \cos n\theta + i \sin n\theta$.
2. $(\cos\theta + i \sin\theta)" = \cos n\theta - i \sin n\theta$
3. $(\cos\theta + i \sin\theta)^{-"} = \cos n\theta - i \sin n\theta$.

1.17 CUBE ROOTS OF UNITY

The three cube roots of unity are 1, w and w^2, *i.e.,* $x = 1$;

$$x = \frac{1+i\sqrt{3}}{2} = w,\ x = \frac{-1-i\sqrt{3}}{2} = w^2.$$

1.18 PROPERTIES OF CUBE ROOTS OF UNITY

(i) $1 + w + w^2 = o$.

(ii) $w^3 = 1$.

(iii) $w^{3n} = 1,\ w^{3+1} = w,\ w^{3n+2} = w^2$

(iv) $a + bw + cw^2 = o \Rightarrow a = b = c$, if a, b, c, are real.

1.19 IMPORTANT RELATIONS IN COMPLEX NUMBERS

(i) $x^2 + y^2 = (x + yi)\ (x - yi)$

(ii) $x^3 + y^3 = (x + y)\ (x + wy)\ (x + w^2y)$.

(iii) $x^3 - y^3 = (x - y)\ (x - wy)\ (x - w^2y)$

(iv) $x^2 + y^2 + z^3 - 3xyz = (x + y + y)\ (x + wy + w^2z)\ (x + w^2y + wz)$.

(v) $x^2 + y^2 + z^2 - xy - yz - yz - zx = (x + wx + w^2z\ (z + w^2y + wz)$.

(vi) $1 = \cos 0 + i \sin 0$; $w = \cos (2\pi/3) + i \sin (2\pi/3)$

$$w^2 = \cos\left(-\frac{2\pi}{3}\right) + i \sin\left[-\frac{2\pi}{3}\right] = \bar{w}$$

$| w | = | w^2 | = 1.$

(vii) $(1 - w + w^2)(1 - w^2 + w^4)(1 - w^4 + w^8)$. –2n factors $= 2^{2n}$

(viii) Euler's theorem : $e^{i\theta} = \cos\theta + i \sin\theta$.

(ix) Exponential representation of z : $z = re^{i\theta}$

(x) nth roots of unity: the n roots are $z = 1$,

$$-e\ i\ (2p/n)\ i\ \text{-----}\ e\ \frac{i2(N-1)\pi}{n}\ \frac{i2\pi k}{n}$$

$z = e$, $k = 0, 1, 2, \ldots\ (\infty - 1)$

$$\Rightarrow z = \cos\left(\frac{2\pi k}{n}\right) + i \sin\left(\frac{2\pi k}{n}\right),\ k = 0, 1, 2\ \text{-----}\ (n-1)$$

1.20 PROPERTIES

Property I.

(i) $| z_1 + z_2 |^2 = | z_1 |^2 + | z_2 |^2 + 2\text{Re}(z_1 \bar{z}_2)$

(ii) $| z_1 - z_2 |^2 = | z_1 |^2 + | z_2 |^2 + 2\text{Re}(z_1 \bar{z}_2)$

(iii) $| z_1 + z_2 |^2 + | z_1 - z_2 |^2 = 2(| z_1 |^2 + | z_2 |^2)$.

Property II.

$| z_1 z_2 | = | z_1 | | z_2 |$.

In Words: *The absolute value of the product of two complex numbers is equal to the product of the absolute values of the numbers.*

Property III. Triangle Inequality.

$| z_1 + z_2 | + | z_1 | + | z_2 |$.

In Words : *The modulus of the sum of two complex numbers cannot be greater than the sum of their moduli.*

Proof:

Let $z_1 = x_1 + iy_1$ and $z_2 = x_2 + iy_2$

so that $| z_1 | = \sqrt{x_1^2 + y_1^2}$ and $| z_2 |\ \sqrt{x_2^2 + y_2^2}$.

Now $z_1 + z_2 = (x_1 + iy_1) + (x_2 + iy_2) = (x_1 + y_1) + i(y_1 + y_2)$

so that $|z_1 + z_2| = \sqrt{(x_1 + x_2)^2 + (y_1 + y_2)^2}$.

Now $|z_1 + z_2| \le |z_1| + |z_2|$ holds true

$\Leftrightarrow \sqrt{(x_1 + x_2)^2 + (y_1 + y_2)^2} \le \sqrt{x_1^2 + y_1^2} + \sqrt{x_2^2 + y_2^2}$

$\Leftrightarrow (x_1 + x_2)^2 + (y_1 + y_2)^2 \le (x_1^2 + y_1^2) + (x_2^2 + y_2^2)$

$+ 2\sqrt{x_1^2 + y_1^2}\sqrt{x_2^2 + y_2^2}$ [Squaring]

$\Leftrightarrow 2x_1 x_2 + 2y_1 y_2 \le + 2\sqrt{x_1^2 + y_1^2}\sqrt{x_2^2 + y_2^2}$

$\Leftrightarrow x_1 x_2 + y_1 y_2 \le \sqrt{x_1^2 + y_1^2}\sqrt{x_2^2 + y_2^2}$

$\Leftrightarrow (x_1 x_2 + y_1 y_2)^2 \le (x_1^2 + y_1^2)(x_2^2 + y_2^2)$ [Squaring]

$\Leftrightarrow x_1^2 x_2^2 + y_1^2 y_2^2 + 2x_1x_2y_1y_2 \le x_1^2 x_2^2 + x_1^2 y_2^2 + x_2^2 y_1^2 + y_1^2 y_2^2$

$\Leftrightarrow 0 \le x_1^2 y_2^2 + x_2^2 y_1^2 - 2x_1x_2y_1y_2$

$\Leftrightarrow 0 \le (x_1 y_2 - x_2 y_2)^2$.

which is true. [$\because$ A perfect sq. of a real number is never –ve]

Extension $|z_1 + z_2 + ... + x_n| \le |z_1| - |z_2| + ... + |z_n|$.

Property IV.

$$\left|\frac{z_1}{z_2}\right| = \left|\frac{z_1}{z_2}\right|, z_2 \ne 0.$$

In Words : *The absolute value of the quotient of two complex numbers (denominator being non-zero) is equal to the quotient of the absolute values of the numbers.*

Property V.

$$|z_1 - z_2| \ge ||z_1| - |z_2||.$$

In Words : *The modulus of the difference of two complex numbers cannot be less than the modulus of the difference of their moduli.*

Property VI.

$$|z_1 - z_2| \ge |z_1| - |z_2|.$$

In Words : *The modulus of the difference of two complex numbers cannot be less than the difference of their moduli.*

Proof:

Here $z_1 - z_2 = (x_1 + iy_1) - (x_2 + iy_2)$

$$= (x_1 - x_2) + i(y_1 - y_2)$$

so that $| z_1 - z_2 | = \sqrt{(x_1 - x_2)^2 + (y_1 - y_2)^2}$

Now $| z_1 - z_2 | \geq | z_1 | - | z_2 |$ holds true

$\Leftrightarrow \quad \sqrt{(x_1 - x_2)^2 + (y_1 - y_2)^2} \geq \sqrt{x_1^2 + y_1^2} - \sqrt{x_2^2 + y_2^2}$

$\Leftrightarrow \quad (x_1 - x_2)^2 + (y_1 - y_2)^2 \geq (x_1^2 + y_1^2)$

$+ (x_2^2 + y_2^2) - 2\sqrt{x_1^2 + y_1^2}\sqrt{x_2^2 + y_2^2}$ [Squaring]

$\Leftrightarrow - 2x_1x_2 - 2y_1y_2 \geq - 2\sqrt{x_1^2 + y_1^2}\sqrt{x_2^2 + y_2^2}$

$\Leftrightarrow x_1x_2 + y_1y_2 \leq \sqrt{x_1^2 + y_1^2}\sqrt{x_2^2 + y_2^2}$ [Dividing by – 2]

$\Leftrightarrow (x_1x_2 + y_1y_2)^2 \leq (x_1^2 + y_1^2)(x_2^2 + y_2^2)$ [Squaring]

$\Leftrightarrow \quad x_1^2x_2^2 + y_1^2y_2^2 + 2x_1x_2y_1y_2 \leq x_1 x_2^2 + x_1^2 y_2^2 + x_2^2 y_1^2 + y_1^2y_2^2$

$\Leftrightarrow 0 \leq x_1^2 y_2^2 + x_2^2 y_1^2 - 2x_1x_2y_1y_2$

$\Leftrightarrow 0 \leq (x_1 y_2 - x_2 y_1)^2$, which is true. [$\because$ A perfect sq. of a real number is never – ve]

1.21 DE-MOIVRE'S THEOREM

(i) If n is any integer, then

$(\cos\theta + i\sin\theta)^n = \cos n\theta + i\sin n\theta.$

(ii) If n is a rational number, then the value or one of the values of $(\cos\theta + i\sin\theta)^n$ is $\cos n\theta + i\sin n\theta$.

(iii) If $z = r(\cos\theta + i\sin\theta)$ and n is a positive integer, then

$$z^{1/n} = r^{1/n}\left[\cos\left(\frac{2k\pi + \theta}{n}\right) + i\sin\left(\frac{2k\pi + \theta}{n}\right)\right],$$

where $k = 0, 1, 2, \ldots, n - 1$.

$\cos\theta + i\sin\theta$ is denoted by cis θ or $e^{i\theta}$.

Thus cis $\theta = e^{i\theta} = \cos\theta + i\sin\theta$.

Note: $| e^{i\theta} | = 1$.

nth Roots of Unity: Let x be a nth root of unity.

Then $x^n = 1 = \cos(2r\pi) + i\sin(2r\pi);\ r \in I$

$\Rightarrow \quad x = [\cos(2r\pi) + i\sin(2r\pi)]^{1/n};\ r \in I$

$$= \cos \frac{2r\pi}{n} + i \sin \frac{2r\pi}{n};$$

$r = 0, 1, 2, \ldots, n - 1$ [De-Moivre's Theorem]

If $\omega = \cos \frac{2\pi}{n} + i \sin \frac{2\pi}{n}$,

then nth roots of unity are $1, \omega, \omega^2, \ldots \omega^{n-1}$.

Result: *Sum of nth roots of unity is zero.*

$$1 + \omega + \omega^2 + \ldots + \omega^{n-1} = \frac{1 - \omega^n}{1 - \omega}$$

$$= \frac{1}{1-\omega}\left[1 - \left\{\cos\left(\frac{2\pi}{n}\right) + i\sin\left(\frac{2\pi}{n}\right)\right\}^n\right]$$

$$= \frac{1}{1-\omega}\,[1 - \{\cos(2\pi) + i\sin(2\pi)\}]$$ [De-Moivre's Theorem]

$$= \frac{1}{1-\omega}[1 - (1 + i(0)] = 0.$$

SOLVED EXAMPLES

Example 1:

Find the least positive integer n for which $\left(\frac{1+i}{1-i}\right)^n = 1$.

Solution:

$$\frac{1+i}{1-i} = \frac{1+i}{1-i} \times \frac{1+i}{1+i} = \frac{(1+i)^2}{1-i^2} = \frac{1+2i+i^2}{1-i^2} = \frac{1+2i-1}{1+1} = \frac{2i}{2} = i$$

$$\therefore \left(\frac{1+i}{1-i}\right)^n = 1 \Rightarrow i^n = 1$$

Hence the least value of n = 4. [$\because i^4 = 1$]

Example 2:

If $\frac{a+ib}{c+id} = x + iy$, *show that*

$$x^2 + y^2 = \frac{a^2 + b^2}{c^2 + d^2}.$$

Solution:

We have $x + iy = \frac{a+ib}{c+id}$

$$= \frac{a + ib}{c + id} \times \frac{c - id}{c - id}$$

$$= \frac{ac - iad + ibc - i^2bd}{c^2 - (id)^2}$$

$$= \frac{(ac + bd) + i(bc - ad)}{c^2 - d^2} \qquad [\because i^2 = -1]$$

$$= \frac{ac + bd}{c^2 + d^2} + i\frac{bc - ad}{c^2 + d^2}$$

$$\therefore \qquad x = \frac{ac + bd}{c^2 + d^2} \text{ and } y = \frac{bc - ad}{c^2 + d^2}$$

$$\therefore \qquad x^2 + y^2 = \frac{(ac + bd)^2}{(c^2 + d^2)^2} + \frac{(bc - ad)^2}{(c^2 + d^2)^2}$$

$$= \frac{a^2c^2 + b^2d^2 + 2abcd + b^2c^2 + a^2d^2 - 2abcd}{(c^2 + d^2)^2}$$

$$= \frac{(a^2c^2 + b^2d^2) + (b^2d^2 + a^2d^2)}{(c^2 + d^2)^2}$$

$$= \frac{c^2(a^2 + b^2) + d^2(b^2 + a^2)}{(c^2 + d^2)^2}$$

$$= \frac{(a^2 + b^2)(c^2 + d^2)}{(c^2 + d^2)^2} = \frac{a^2 + b^2}{c^2 + d^2}.$$

Example 3:

For what real values of x and y are the numbers $-3 + ix^2y$ *and* $x^2 + y + 4i$ *conjugate complex?*

Solution:

The conjugate of $x^2 + y + 4i$ is $x^2 + y - 4i$.

By the question, $x^2 + y - 4i = -3 + ix^2y$

Equating real parts, $\quad x^2 + y = -3 \qquad \ldots(1)$

Equating imag. parts $\quad -4 = x^2y \qquad \ldots(2)$

From (2), $\quad x^2 = -\frac{4}{y} \qquad \ldots(3)$

Putting (1) $\quad -\frac{4}{y} + y = -3$

$\Rightarrow \quad -4 + y^2 = -3y$

$\Rightarrow \quad y^2 + 3y - 4 = 0$

$\Rightarrow \quad (y + 4)(y - 1) = 0$

$\Rightarrow \quad y = -4, 1.$

When $y = -4$, then from (3), $x^2 = -\dfrac{4}{-4} = 1$

$\Rightarrow \quad x = \pm 1$

When $y = 1$, then from (3) $x^2 = -4$

$\Rightarrow$ x is non-real.

Hence $x = 1, y = -4$ and $x = -1, y = -4$.

Example 4:

Solve the equation $|z| - z = 1 + 2i$, *where* $z = x + iy$.

Solution:

We have $|z| - z = 1 + 2i$

$$\Rightarrow \quad \sqrt{x^2 + y^2} - (x + iy) = 1 + 2i$$

Equating real parts, $\sqrt{x^2 + y^2} - x = 1$ ...(1)

Equating imag. parts, $-y = 2$ ...(2)

From (2), $y = -2.$

Putting in (1) $\sqrt{x^2 + 4} - x = 1$

$$\Rightarrow \quad \sqrt{x^2 + 4} = 1 + z$$

Squaring, $x^2 + 4 = 1 + 2x + x^2$

$\Rightarrow \quad 4 = 1 + 2x$

$\Rightarrow \quad 2x = 3 \Rightarrow x = \dfrac{3}{2}.$

Hence the solution is $z = \dfrac{3}{2} - 2i.$

Example 5:

Find the real values of x and y if

$$\sqrt{x}\,(i + \sqrt{y}) - 15 = i(8 - \sqrt{y})$$

Solution:

We have $\sqrt{x}\,(i + \sqrt{y}) - 15 = i\,(8 - \sqrt{y})$

$\Rightarrow \quad i\sqrt{x} + \sqrt{xy} - 15 = 8i - \sqrt{y}\,i$

$\Rightarrow \quad (\sqrt{xy} - 15) + i\sqrt{x} = (8 - \sqrt{y})i$

Equating real parts, $\sqrt{xy} - 15 = 0$

$\Rightarrow \quad \sqrt{xy} = 15$ ...(1)

Equating imaginary parts, $\sqrt{x} = 8 - \sqrt{y}$

$\Rightarrow \quad \sqrt{x} + \sqrt{y} = 8$ ...(2)

Squaring (2), $x + y + 2\sqrt{xy} = 64$

$\Rightarrow \quad x + y + 2(15) = 64$ [Using (1)]

$\Rightarrow \quad x + y = 64 - 30 = 34$

$\Rightarrow \quad y = 34 - x$ ...(3)

Squaring (1), $xy = 225$

$\Rightarrow \quad x(34 - x) = 225$ [Using (3)]

$\Rightarrow \quad 34x - x^2 = 225$

$\Rightarrow \quad x^2 - 34x + 225 = 0$

$\Rightarrow \quad (x - 9)(x - 25) = 0$

$\Rightarrow \quad x = 9$ or 25.

When $x = 9$, then $y = 34 - 9 = 25$.

When $x = 25$, then $y = 34 - 25 = 9$.

Hence $\left.\begin{matrix} x = 9 \\ y = 25 \end{matrix}\right\}, \quad \left.\begin{matrix} x = 25 \\ y = 9 \end{matrix}\right\}.$

Example 6:

Let z_1, z_2 be any two complex numbers. Then, show that

$$|z_1 + \sqrt{z_1^2 - z_2^2}| + |z_1 - \sqrt{z_1^2 - z_2^2}| = |z_1 + z_2| + z_1 - z_2|.$$

Solution:

$(|z| + \sqrt{z_1^2 - z_2^2}| + |z_1 - \sqrt{z_1^2 - z_2^2}|)^2$

$$= | z_1 + \sqrt{z_1^2 - z_2^2} |^2 + | z_1 - \sqrt{z_1^2 - z_2^2} |^2$$

$$+ 2 | z_1 + \sqrt{z_1^2 - z_2^2} | | z_1 - \sqrt{z_1^2 - z_2^2} |$$

$$= 2(| z_1 |^2 + | \sqrt{z_1^2 - z_2^2} |^2) + 2 | (z_1 + \sqrt{z_1^2 - z_2^2}) (z_1 - \sqrt{z_1^2 - z_2^2}) |$$

$$= 2(| z_1 |^2 + | z_1^2 - z_2^2 |) + 2 | z_1^2 - (z_1^2 - z_1^2) |$$

$$= 2(| z_1 |^2 + | z_1^2 - z_2^2 |) + 2 | z_2^2 |$$

$$= 2(| z_1 |^2 + | z_2 |^2) + 2 | z_1^2 - z_2^2 |$$

$$= | z_1 + z_2 |^2 + | z_1 - z_2 |^2 + 2 | z_1 + z_2 | | z_1 - z_1 |$$

$$= (| z_1 + z_2 | + | z_1 - z_2 |)^2$$

Hence $| z_1 + \sqrt{z_1^2 - z_2^2} | + | z_1 - \sqrt{z_1^2 - z_2^2} |$

$$= | z_1 + z_2 | + | z_1 - z_2 |.$$

Example 7:

If 1, ω, ω^2 are three cube roots of unity, prove that
$(1 + \omega) (1 + \omega^2) (1 + \omega^4) (1 + \omega^8)$ ... to 2n factors = 1.

Solution:

L.H.S. = $(1 + \omega)(1 + \omega^2)(1 + \omega^4)(1 + \omega^8)$ ... to 2n factors.

$= (1 + \omega)(1 + \omega^2)(1 + \omega^3.\omega)(1 + \omega^6 .\omega^2)$ ... to 2n factors

$= (1 + \omega)(1 + \omega^2)(1 + \omega)(1 + \omega^2)$ ... to 2n factors

[$\because \omega^3 = \omega^6 = \ldots = 1$]

$= [(1 + \omega)(1 + \omega)$ ... to n factors]

$= [(1 + \omega^2)(1 + \omega^2)$ ... to n factors] [Note this step]

$= (1 + \omega)^n(1 + \omega^2)^n = (1 + \omega + \omega^2 + \omega^3)^n$

$= (0 + 1)^n$ [$\because 1 + \omega + \omega^2 = 0$ and $\omega^3 = 1$]

$= 1.$

Example 8:

If 1, ω, ω^2 are three cube roots of unity and $x = a + b$, $y = a\omega + b\omega^2$, $z = a\omega^2 + b\omega$, then prove that $x^2 + y^2 + z^2 = 6ab$.

Solution:

We have $x = a + b$, $y = a\omega + b\omega^2$,

$z = a\omega^2 + b\omega.$

$$\text{L.H.S.} = x^2 + y^2 + z^2$$
$$= (a + b)^2 + (a\omega + b\omega^2)^2 + (a\omega^2 + b\omega)^2$$
$$= a^2 + b^2 + 2ab + a^2\omega^2 + b^2\omega^4 + 2ab\omega^3 + a^2\omega^4 + b^2\omega^2 + 2ab\omega^3$$
$$= a^2 + b^2 + 2ab + a^2\omega^2 + b^2\omega + 2ab\ (1) + a^2\omega + b^2\omega^2 + 2ab\ (1)$$
$$[\because\ \omega^2 = 1 \text{ and } \omega^4 = \omega^3.\omega = 1.\ \omega = \omega]$$
$$= a^2(1 + \omega + \omega^2) + b^2(1 + \omega + \omega^2) + 6ab$$
$$= a^2(0)\ b^2(0) + 6ab \quad [\because\ 1 + \omega + \omega^2 = 0]$$
$$= 6ab = \text{R.H.S.}$$

Example 9:

Prove that $\left(\dfrac{-1 + i\sqrt{3}}{2}\right)^n + \left(\dfrac{-1 - i\sqrt{3}}{2}\right)^n$ *is equal to 2 if n be a multiple of 3 and is equal to – 1 if n any other integer.*

Or

If 1, ω, ω^2 are the cube roots of unity, prove that $\omega^n + \omega^{2n} = 2$ or -1 according as n is a multiple of 3 or any other integer.

Solution:

Three cases arise:

Case I: *When n = 3k, where k is any +ve integer.*

$$\left(\frac{-1 + i\sqrt{3}}{2}\right)^n + \left(\frac{-1 - i\sqrt{3}}{2}\right)^n = (\omega)^{3k} + (\omega^2)^{3k}$$
$$= (\omega^3)^k + (\omega^3)^{2k}$$
$$= (1)^k + (1)^{2k} \qquad [\because\ \omega^3 = 1]$$
$$= 1 + 1 = 2.$$

Case II: *When n = 3k + 1, where k is any +ve integer.*

$$\left(\frac{-1 + i\sqrt{3}}{2}\right)^n + \left(\frac{-1 - i\sqrt{3}}{2}\right)^n = (\omega)^{3k+1} + (\omega^2)^{3k+1}$$
$$= (\omega)^{3k}\ \omega + (\omega)^{6k}\ \omega^2$$
$$= (\omega^3)^3\omega + (\omega^3)^{2k}\ \omega^2$$
$$= (1)^k\omega + (1)^{2k}\ \omega^2 \qquad [\because\ \omega^3 = 1]$$
$$= \omega + \omega^2 = \omega^2 + \omega = -1. \qquad [\because\ 1 + \omega + \omega^2 = 0]$$

Case III: *When n = 3k + 2, where k is any +ve integer.*

$$\left(\frac{-1+i\sqrt{3}}{2}\right)^n + \left(\frac{-1-i\sqrt{3}}{2}\right)^n = (\omega)^{3k+2} + (\omega^2)^{3k+2}$$

$$= (\omega^3)^k \omega^2 + (\omega)^{6k} \omega^4$$

$$= (\omega^3)^k\omega^2 + (\omega^3)^{2k+1} . \omega$$

$$= (1)^k(\omega) + (1)^{2k+1} (\omega) \qquad [\because \omega^3 = 1]$$

$$= \omega^2 + \omega = -1. \qquad [\because 1 = \omega + \omega^2 = 0]$$

Solution 10:

If 1, ω, ω² be cube roots of unity, prove that $(x - y)(x\omega - y)(x\omega^2 - y) = x^3 - y^3$.

Solution:

$$\text{L.H.S.} = (x - y)(x\omega - y)(x\omega^2 - y)$$

$$= (x - y)[x^2\omega^3 - xy\omega - xy\omega^2 + y^2]$$

$$= (x - y) . [x^2\omega^3 - (\omega + \omega^2)xy + y^2]$$

$$= (x - y) . [x^2 . 1 - (-1)xy + y^2) \quad [\because \omega^2 = 1 \text{ and } 1 + \omega + \omega^2 = 0]$$

$$= (x - y)(x^2 + xy + y^2) = x^3 - y^3$$

$$= \text{R.H.S.}$$

Example 11:

If 1, $\omega_1, \omega_2, \ldots, \omega_{n-1}$ *are the nth roots of unity, then show that* $(1 - \omega_1)(1 - \omega_2)(1 - \omega_3) \ldots (1 - \omega_{n-1}) = n$.

Solution:

Since $1, \omega_1, \omega_2, \ldots, \omega_{n-1}$ are roots of $x^n = 1$,

$$\therefore \quad x^n - 1 = (x - 1)(x - \omega_1)(x - \omega_2) \ldots (x - \omega_{n-1})$$

$$\Rightarrow (x - 1)(x^{n-1} + x^{n-2} + \ldots + x + 1)$$

$$= (x - 1)(x - \omega_1)(x - \omega_2) \ldots (x - \omega_{n-1})$$

$$\Rightarrow x^{n-1} + x^{n-2} + \ldots + x + 1 = (x - \omega_1)(x - \omega_2) \ldots (x - \omega_{n-1})$$

Putting $x = 1$, $n = (1 - \omega_1)(1 - \omega_2) \ldots (1 - \omega_{n-1})$, which is true.

Example 12:

Evaluate $\dfrac{(-1 + i\sqrt{3})^{15}}{(1 - i)^{20}} + \dfrac{(-1 - i\sqrt{3})^{15}}{(1 + i)^{20}}$.

Solution:

Since $\dfrac{-1+i\sqrt{3}}{2} = \omega$ and $\dfrac{-1-i\sqrt{3}}{2} = \omega^2$.

$\therefore -1 + i\sqrt{3} = 2\omega$ and $-1 + i\sqrt{3} = 2\omega^2$.

$$\therefore \text{ Given expression } = \frac{2(\omega)^{15}}{(1-i)^{20}} + \frac{2(\omega^2)^{15}}{(1+i)^{20}}$$

$$\begin{aligned}
\text{But} \quad (1-i)^{20} &= ((1-i)^2)^{10} = (1 + i^2 - 2i)^{10} \\
&= (1 - 1 - 2i)^{10} = (-2i)^{10} \\
&= 2^{10} i^{10} = 2^{10}(i^2)^5 \\
&= 2^{10}(-1)^5 = -2^{10}
\end{aligned}$$

and $(1+i)^{20} = -2^{10}$ (Similarly)

$$\therefore \text{ Given expression } = -\frac{1}{2^{10}}\left[(2\omega)^{15} + (2\omega^2)^{15}\right]$$

$$= -\frac{1}{2^{10}}\left[2^{15} + 2^{15}\right] \qquad [\because \omega^{15} = \omega^{30} = 1]$$

$$= -\frac{1}{2^{10}}(2 \,.\, 2^{15}) = -2^6 = -64.$$

Example 13:

If $z = \left(\dfrac{\sqrt{3}}{2} + \dfrac{i}{2}\right)^{107} + \left(\dfrac{\sqrt{3}}{2} - \dfrac{i}{2}\right)^{107}$, *then show that Im (z) = 0.*

Solution:

$$\text{We have } z = \left(\frac{\sqrt{3}}{2} + \frac{i}{2}\right)^{107} + \left(\frac{\sqrt{3}}{2} - \frac{i}{2}\right)^{107}$$

$$\therefore \quad \bar{z} = \left(\frac{\sqrt{3}}{2} + \frac{i}{2}\right)^{107} + \left(\frac{\sqrt{3}}{2} - \frac{i}{2}\right)^{107} = z$$

$\Rightarrow$ Im (z) = 0.

Example 14:

Let z_1, z_2, z_3 *be non-real complex numbers and* $z_1 \neq z_2$.

If $\begin{vmatrix} |z_1| & |z_2| & |z_3| \\ |z_2| & |z_3| & |z_1| \\ |z_3| & |z_1| & |z_2| \end{vmatrix} = 0$, *then prove that (i)* z_1, z_2, z_3 *lie on a circle with centre at the origin.*

(ii) amp. $\left(\frac{z_1}{z_2}\right)$ = amp. $\left(\frac{z_3 - z_1}{z_2 - z_1}\right)^2$.

Solution:

Let $z_1 = r_1 (\cos \alpha_1 + i \sin \alpha_1)$, $z_2 = r_2 (\cos \alpha_2 + i \sin \alpha_2)$ and $z_3 = r_3 (\cos \alpha_3 + i \sin \alpha_3)$.

By the question, $\begin{vmatrix} r_1 & r_2 & r_3 \\ r_2 & r_3 & r_1 \\ r_3 & r_1 & r_2 \end{vmatrix} = 0$

$\Rightarrow \quad \begin{vmatrix} r_1 + r_2 + r_3 & r_2 & r_3 \\ r_1 + r_2 + r_3 & r_3 & r_1 \\ r_1 + r_2 + r_3 & r_1 & r_2 \end{vmatrix} = 0$ [Operating $C_1 \to C_1 + C_2 + C_3$]

$\Rightarrow \quad (r_1 + r_2 + r_3) \begin{vmatrix} 1 & r_2 & r_3 \\ 1 & r_3 & r_1 \\ 1 & r_1 & r_2 \end{vmatrix} = 0$

$\Rightarrow \quad (r_1 + r_2 + r_3) \begin{vmatrix} 1 & r_2 & r_3 \\ 0 & r_3 - r_2 & r_1 - r_3 \\ 0 & r_1 - r_2 & r_2 - r_3 \end{vmatrix} = 0$

[Operating $R_2 \to R_2 - R_1$ and $R_3 \to R_3 - R_1$]

$\Rightarrow \quad (r_1 + r_2 + r_3) [(r_3 - r_2)(r_2 - r_3) - (r_1 - r_2)(r_1 - r_2)] = 0$

$\Rightarrow \quad -(r_1 + r_2 + r_3)(r_1^2 + r_2^2 + r_3^3 - r_1 r_2 - r_2 r_3 - r_3 r_1) = 0$

$\Rightarrow \quad \frac{1}{2}(r_1 + r_2 + r_3) [(r_1 - r_2)^2 + (r_2 - r_3)^2 + (r_3 - r_1)^2] = 0$

But $r_1 + r_2 + r_3 \neq 0$. [$\because r_1, r_2, r_3$ are all + ve]

$\therefore r_1 - r_2 = r_2 - r_3 = r_3 - r_1 = 0$

$\Rightarrow \quad r_1 = r_2 = r_3$

$\Rightarrow \quad z_1, z_2, z_3$ are three complex numbers whose distances from the origin and equal

$\Rightarrow$ z_1, z_2, z_3 lie on a circle with centre at the origin.

Let the point representing the complex numbers z_1, z_2, z_3 be A, B, C respectively.

Also amp. $\left(\frac{z_3}{z_2}\right)$ = angle which [CB] makes at the origin and amp. $\left(\frac{z_3 - z_1}{z_2 - z_1}\right)^2 = 2$ amp. $\left(\frac{z_3 - z_1}{z_2 - z_1}\right) = 2\angle CAB$.

$$\text{Hence amp.} \left(\frac{z_3}{z_2}\right) = 2 \text{ amp.} \left(\frac{z_3 - z_1}{z_2 - z_1}\right)$$

$$= \text{amp.} \left(\frac{z_3 - z_1}{z_2 - z_1}\right)^2.$$

Example 15:

Given $\prod_{k=1}^{n}(P_k + iQ) = R + iS$, *prove that*

$$\sum_{k=1}^{n} \tan^{-1}\frac{Q_k}{P_k} = \tan^{-1}\frac{S}{R} + n\pi.$$

Solution:

Let $P_1 + iQ_1 = r_1(\cos\theta_1 + i\sin\theta_1)$

$P_2 + iQ_2 = r_2(\cos\theta_2 + i\sin\theta_2)$

....

$P_n + iQ_n = r_n(\cos\theta_n + i\sin\theta_n)$

Also $R + iS = r(\cos\phi + i\sin\phi)$

Now $\prod_{k=1}^{n}(P_k + iQ) = R + iS$

$\Rightarrow$ $r_1 r_2 \dots r_n [\cos\theta_1 + i\sin\theta_1][\cos\theta_2 + i\sin\theta_2]$
$\dots [\cos\theta_n + i\sin\theta_n] = r(\cos\phi + i\sin\phi)$

$\Rightarrow$ $r_1 r_2 \dots r_n [\cos(\theta_1 + \theta_2 + \dots + \theta_n) + i\sin(\theta_1$
$+ \theta_2 + \dots \theta_n)] = r(\cos\phi + i\sin\phi)$

Comparing, $r_1 r_2 \dots r_n = r$, $\cos(\theta_1 + \theta_2 + \dots + \theta_n)$
$= \cos\phi$ and $\sin(\theta_1 + \theta_2 + \dots + \theta_n) = \sin\phi$

$$\therefore \quad \frac{\sin(\theta_1 + \theta_2 + \dots + \theta_n)}{\cos(\theta_1 + \theta_2 + \dots + \theta_n)} = \frac{\sin\phi}{\cos\phi}$$

$\Rightarrow \quad \tan(\theta_1 + \theta_2 + \ldots + \theta_n) = \tan\phi$

$\Rightarrow \quad \theta_1 + \theta_2 + \ldots + \theta_n = n\pi + \phi.$

But $\quad \theta_1 = \text{amp}(P_1 + iQ_1) = \tan^{-1}\dfrac{Q_1}{P_1}$

$\theta_2 = \text{amp}(P_2 + iQ_2) = \tan^{-1}\dfrac{Q_2}{P_2}$

....

$\phi = \text{amp}(R + S) = \tan^{-1}\dfrac{S}{R}.$ **Hence the result.**

Example 16:

If the vertices of a triangle ABC represent z_1, z_2, z_3 respectively, show that the orthocentre represents

$$\frac{(a\sec A)z_1 + (b\sec B)z_2 + (c\sec C)z_3}{a\sec A + b\sec B + c\sec C}$$

Solution:

Let $A(z_1)$, $B(z_2)$ and $C(z_3)$ be the vertices of ΔABC. Let AD, BE, CF be the altitudes which meet at H, the orthocentre.

We have $\quad \dfrac{AD}{AB} = \cos B$

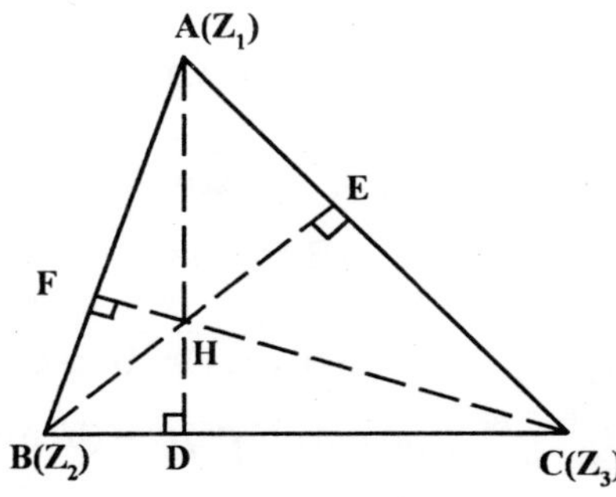

$\Rightarrow \quad \dfrac{BD}{c} = \cos B$

$\Rightarrow \quad BD = c\cos B$

and $\quad \dfrac{DC}{AC} = \cos C$

$\Rightarrow \quad \dfrac{DC}{b} = \cos C$

$\Rightarrow \quad DC = b\cos C.$

Thus D divides the segment [BC] in the ratio $c\cos B : b\cos C$

$\therefore$ D represents $\quad \dfrac{(c\cos B)z_3 + (b\cos C)z_2}{c\cos B + b\cos C}$

i.e., $\quad \dfrac{(b\cos C)z_2 + (a\cos B)z_3}{a}$

In ΔABH, by Sine rule,

$$\frac{AH}{\sin(90° - A)} = \frac{BH}{\sin(90° - B)} = \frac{AB}{\sin(A + B)}$$

$$\Rightarrow \quad \frac{AH}{\cos A} = \frac{BH}{\cos B} = \frac{c}{\sin C} = 2R$$

$\therefore$ AH = 2R cos A and BH = 2R cos B.

In ΔBDH, $\quad \dfrac{HD}{BH} = \sin(90° - C)$

$$\Rightarrow \quad HD = 2R \cos B \cos C$$

$$\therefore \quad AH : HD = 2R \cos A : 2R \cos B \cos C$$

$$= \cos A : \cos B \cos C$$

$\therefore$ H divides the segment [AD] in the ratio cos A : cos B cos C

$\therefore$ H represents

$$\frac{\cos A\left[\dfrac{(b\cos C)z_2 + (c\cos B)z_3}{a}\right] + (\cos B\cos C)z_1}{\cos A + \cos B\cos C}$$

i.e., $$\frac{(a\cos B\cos C)z_1 + (b\cos C\cos A)z_2 + (c\cos A\cos B)z_3}{a[\cos A + \cos B\cos C]}$$

i.e., $$\frac{(a\cos B\cos C)z_1 + (b\cos C\cos A)z_2 + (c\cos A\cos B)z_3}{a\cos A + a\cos B\cos C}$$

i.e., $$\frac{(a\cos B\cos C)z_1 + (b\cos C\cos A)z_2 + (c\cos A\cos B)z_3}{a\cos B\cos C + (b\cos C + c\cos B)\cos A}$$

i.e., $$\frac{(a\cos B\cos C)z_1 + (b\cos C\cos A)z_2 + (c\cos A\cos B)z_3}{a\cos B\cos C + b\cos C + \cos A + c\cos A\cos B}$$

i.e., $$\frac{(a\sec A)z_1 + (b\sec B)z_2 + (c\sec C)z_3}{a\sec A + b\sec B + c\sec C}$$

[Dividing numerator and denominator by cos A cos B cos C]

Example 17:

ABCD is a rhombus described in the clockwise direction in the Argand diagram. Suppose that the vertices A, B, C, D are given by z_1, z_2, z_3, z_4 *respectively and* $\angle CBA = \dfrac{2\pi}{3}$.

Show that $2\sqrt{3}\, z_2 = (\sqrt{3} - i)z_1 + (\sqrt{3} + i)z_3$

and $2\sqrt{3}\, z_4 = (\sqrt{3} + i)z_1 + (\sqrt{3} - i)z_3.$

Solution:

Given $\angle \text{CBA} = \dfrac{2\pi}{3}$

$$\therefore \quad \frac{z_3 - z_2}{z_1 - z_2} = \frac{|z_3 - z_2|}{|z_1 - z_2|}\left[\cos\left(\frac{2\pi}{3}\right) + i\sin\left(\frac{2\pi}{3}\right)\right]$$

$$\Rightarrow \quad \frac{z_3 - z_2}{z_1 - z_2} = -\frac{1}{2} + i\frac{\sqrt{3}}{2}$$

[$\because$ $|z_3 - z_2| = |z_1 - z_2|$ as ABCD is a rhombus]

$$\Rightarrow \quad z_3 - z_2 = \left(-\frac{1}{2} + \frac{i\sqrt{3}}{2}\right)(z_1 - z_2)$$

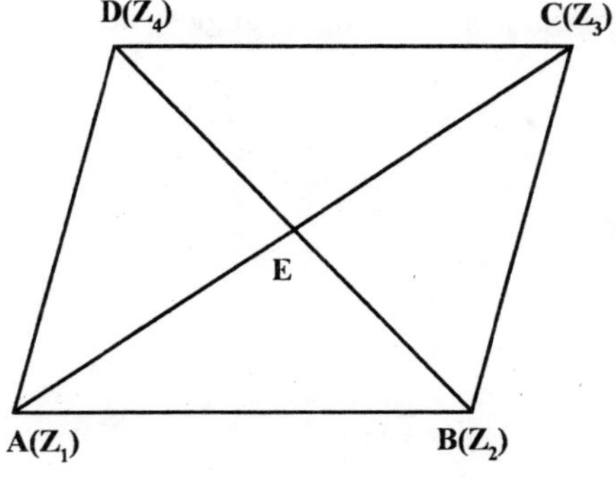

$$\Rightarrow \quad z_3 + \left(\frac{1}{2} - \frac{i\sqrt{3}}{2}\right)z_1 = \left(\frac{3}{2} - \frac{i\sqrt{3}}{2}\right)z_2$$

$$\Rightarrow \quad \sqrt{3}z_2 = \frac{2}{\sqrt{3} - i}z_3 + \frac{1 - i\sqrt{3}}{\sqrt{3} - i}z_1$$

$$\Rightarrow \quad 2\sqrt{3}\, z_2 = (\sqrt{3} + i)z_3 + \frac{1}{2}(\sqrt{3} + i)(1 - i\sqrt{3})z_1$$

$$= (\sqrt{3} + i)z_3 + \frac{1}{2}(\sqrt{3} + \sqrt{3} + i - 3i)z_1$$

$$= (\sqrt{3} + i)z_3 + (\sqrt{3} - i)z_1 \qquad \ldots(1)$$

Since the diagonals of a rhombus bisect each other,

$$\therefore \frac{z_1 + z_3}{2} = \frac{z_2 + z_4}{2} \Rightarrow z_1 + z_3 = z_2 + z_4$$

$\Rightarrow \quad z_4 = z_1 + z_3 - z_2$

$\Rightarrow \quad 2\sqrt{3}\, z_4 = 2\sqrt{3}\,(z_1 + z_3) - 2\sqrt{3}\, z_2$

$= 2\sqrt{3}\,(z_1 + z_3) - [(\sqrt{3} + i)z_3 + (\sqrt{3} - i)z_1]$ [Using (1)]

$= (\sqrt{3} + i)z_1 + (\sqrt{3} - i)z_3$ ...(2)

Hence (1) and (2) are the reqd. results.

Example 18:

Show that the circumcentre of the triangle whose vertices are given by the complex numbers z_1, z_2, z_3 is given by

$$z = \frac{\Sigma z_1 \bar{z}_2 (z_2 - z_3)}{\Sigma \bar{z}_1 (z_2 - z_3)}.$$

Solution:

Let O(z) be the circumcentre of the triangle whose vertices are $A(z_1)$, $B(z_2)$ and $C(z_3)$.

By def., $|z - z_1| = |z - z_2| = |z - z_3|$.

Now $\quad |z - z_1| = |z - z_2|$

$\Rightarrow \quad |z - z_1|^2 = |z - z_2|^2$

$\Rightarrow \quad |z|^2 + |z_1|^2 - \bar{z} z_1 - z\bar{z}_1 = |z|^2 + |z_2|^2 - \bar{z} z_2 - z\bar{z}_2$

$\Rightarrow \quad z(\bar{z}_1 - \bar{z}_2) + \bar{z}(z_1 - z_2) = z_1\bar{z}_1 - z_2\bar{z}_2$ ...(1)

Similarly $|z - z_1| = |z - z_3|$ gives

$z(\bar{z}_1 - \bar{z}_3) + \bar{z}(z_1 - z_3) = z_1\bar{z}_1 - z_3\bar{z}_3$ ...(2)

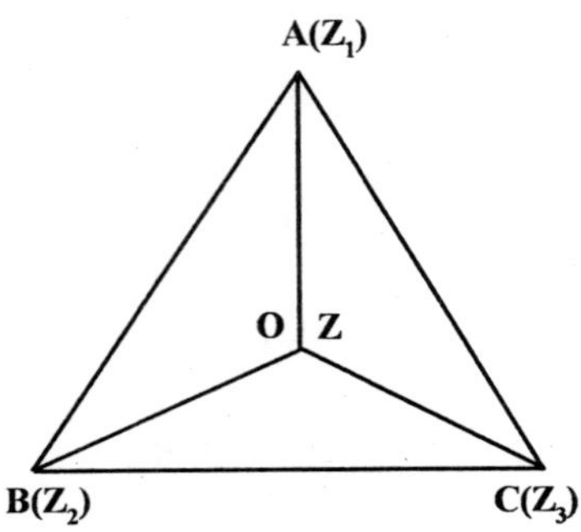

Multiplying (1) by $z_1 - z_3$ and (2) by $z_1 - z_2$ and subtracting, we get

$$z[(\bar{z}_1 - \bar{z}_2)(z_1 - z_2) - (\bar{z}_1 - \bar{z}_3)(z_1 - z_2)]$$
$$= z_1\bar{z}_1(z_1 - z_3) - z_2\bar{z}_2(z_1 - z_3) - z_1\bar{z}_1(z_1 - z_2) + z_3\bar{z}_3 + (z_1 - z_2)$$
$$\Rightarrow \quad z[\bar{z}_1(z_2 - z_3) + \bar{z}_2(z_3 - z_1) + \bar{z}_3(z_1 - z_2)]$$
$$= z_1\bar{z}_1(z_2 - z_3) + z_2\bar{z}_2(z_3 - z_1) + z_3\bar{z}_3(z_1 - z_2).$$

Hence $\quad z = \dfrac{\Sigma z_1\bar{z}_2\,(z_2 - z_3)}{\Sigma \bar{z}_1\,(z_2 - z_3)}.$

Example 19:

Prove that $|z_1 + z_2|^2 + |z_1 - z_2|^2 = 2|z_1|^2 + |z_2|^2$.

Interpret the result geometrically and deduce that

$$|\alpha + \sqrt{\alpha^2 - \beta^2}| + |\alpha - \sqrt{\alpha^2 - \beta^2}| = |\alpha + \beta| + |\alpha - \beta|.$$

where α, β *are complex numbers.*

Solution:

$$|z_1 + z_2|^2 + |z_1 - z_2|^2$$
$$= (z_1 + z_2)\,(\overline{z_1 + z_2}) + (z_1 - z_2)\,(\overline{z_1 - z_2})$$
$$= (z_1 + z_2)\,(\bar{z}_1 + \bar{z}_2) + (z_1 - z_2)\,(\bar{z}_1 - \bar{z}_2)$$
$$= z_1\bar{z}_1 + z_2\bar{z}_1 + z_1\bar{z}_2 + z_2\bar{z}_2 + z_1\bar{z}_1 - z_2\bar{z}_1 - z_1\bar{z}_2 + z_2\bar{z}_2$$
$$= 2z_1\bar{z}_2 + 2z_2\bar{z}_2 = 2|z_1|^2 + 2|z_2|^2 \qquad \ldots(1)$$

which is true.

Geometrical Interpretation

Let A, B be two points which represent two complex numbers z_1, z_2 respectively in the complex plane. Complete the || gm OACB.

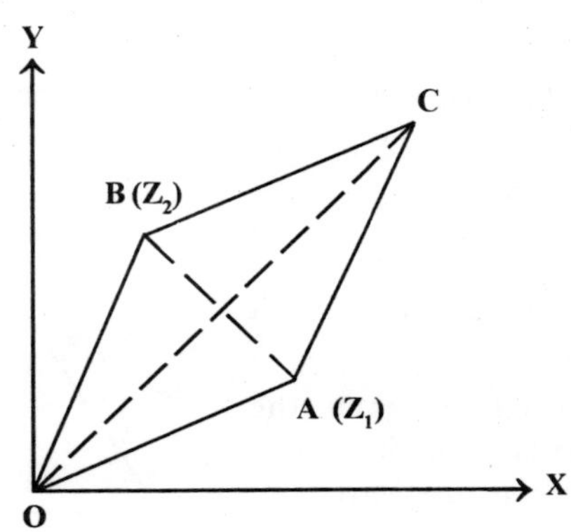

Then $|z_1| = OA$, $|z_2| = OB$,

$|z_1 + z_2| = OC$, $|z_1 - z_2| = AB$.

Thus, the sum of the squares of the diagonals of a parallelogram is equal to the sum of the squares of its sides.

Deduction. Let $z_1 = \alpha + \sqrt{\alpha^2 - \beta^2}$

and $z_2 = \alpha - \sqrt{\alpha^2 - \beta^2}$

$\therefore$ From (1), $|z_1|^2 + |z_2|^2 = \frac{1}{2}[|z_1 + z_2|^2 + |z_1 - z_2|^2]$

$= \frac{1}{2}[|2\alpha|^2 + |2\sqrt{\alpha^2 - \beta^2}|^2]$

$= 2[|\alpha|^2 + |\alpha^2 - \beta^2|]$ ...(2)

Also $(|z_1| + |z_2|)^2 = |z_1|^2 + |z_2|^2 + 2|z_1||z_2|$

$= 2[|\alpha|^2 + |\alpha^2 - \beta^2|] + 2[\beta^2]$ [Using (2)]

$= |\alpha + \beta|^2 + |\alpha - b|^2 + 2|\alpha^2 - \beta^2|$ [Using (1)]

$= (|\alpha + \beta| + |\alpha - \beta|)^2$

$\Rightarrow$ $|z_1| + |z_2| = |\alpha + \beta| + |\alpha - \beta|$.

Hence $|\alpha + \sqrt{\alpha^2 + \beta^2}| + |\alpha - \sqrt{\alpha^2 - \beta^2}| = |\alpha + \beta| + |\alpha - \beta|$.

Example 20:

If z_1, z_2, z_3 are the vertices of an isosceles triangle right angled at the vertex z_2, show that

$$z_1^2 + 2z_2^2 + z_3^2 = 2z_2(z_1 + z_3).$$

Solution:

Let $P(z_1)$, $Q(z_2)$ and $R(z_3)$ be the vertices of the isosceles triangle PQR, right angled at Q.

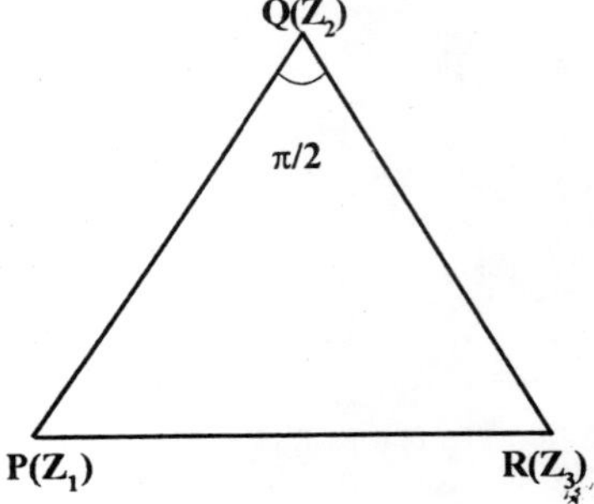

$\therefore \angle PQR = \frac{\pi}{2}$.

Thus, $\arg\left(\frac{z_2 - z_1}{z_2 - z_3}\right) = \frac{\pi}{2}$.

This is possible when the complex number is purely imaginary *i.e.*, its real part must be zero.

Now real part of $\left(\dfrac{z_2 - z_1}{z_2 - z_3}\right) = 0$.

$$\Rightarrow \quad \frac{1}{2}\left[\frac{z_2 - z_1}{z_2 - z_3} + \left(\overline{\frac{z_2 - z_1}{z_2 - z_3}}\right)\right] = 0, \text{ since } \operatorname{Re}(z) = \frac{1}{2}(z + \overline{z})$$

$$\Rightarrow \quad \frac{z_2 - z_1}{z_2 - z_3} = \frac{\overline{z}_2 - \overline{z}_1}{\overline{z}_2 - \overline{z}_3} = 0$$

$$\Rightarrow \quad \frac{z_2 - z_1}{z_2 - z_3} = -\frac{\overline{z}_2 - \overline{z}_1}{\overline{z}_2 - \overline{z}_3}$$

$$\Rightarrow \quad \frac{z_2 - z_1}{\overline{z}_2 - \overline{z}_1} = -\frac{z_2 - z_3}{\overline{z}_2 - \overline{z}_3} \qquad \ldots(1)$$

Since the triangle PQR is isosceles *i.e.*, PQ = QR, we have

$$| z_2 - z_1 | = | z_2 - z_3 |$$

$$\Rightarrow \quad | z_2 - z_1 |^2 = | z_2 - z_3 |^2$$

$$\Rightarrow \quad (z_2 - z_1)(\overline{z}_2 - \overline{z}_1) = (z_2 - z_3)(\overline{z}_2 - \overline{z}_3) \qquad \ldots(2)$$

on using the result $| z |^2 = z\overline{z}$.

Multiplying corresponding sides of (1) and (2), we have

$$(z_2 - z_1)^2 = -(z_2 - z_3)^2$$

$$\Rightarrow \quad z_1^2 + 2z_2^2 + z_3^2 = 2z_2(z_1 + z_2), \text{ which is true.}$$

Example 21:

Let the complex numbers z_1, z_2, z_3 be the vertices of an equilateral triangle. Let z_0 be the circumcentre of the triangle, then prove that

$$z_1^2 + z_2^2 + z_3^2 = 8z_0^2.$$

Solution:

Since Δ ABC is equilateral, we have

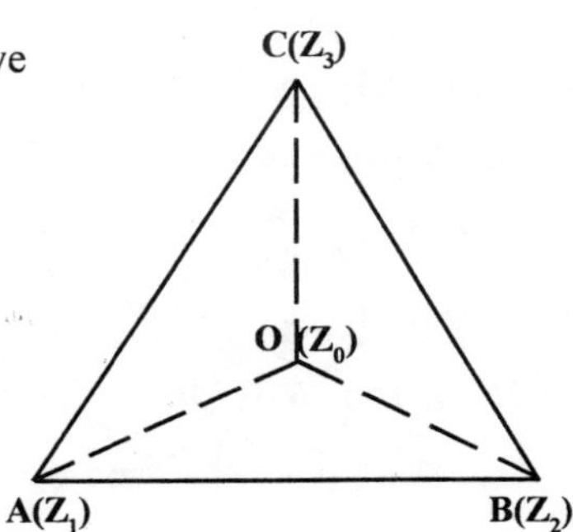

$$AB = BC = CA$$

$$\Rightarrow \quad \frac{AB}{BC} = \frac{BC}{CA} = 1$$

$$\Rightarrow \quad \left|\frac{z_2 - z_1}{z_3 - z_2}\right| = \left|\frac{z_2 - z_3}{z_3 - z_1}\right| = 1$$

$$\Rightarrow \quad \frac{z_2 - z_1}{z_3 - z_2} = \frac{z_2 - z_3}{z_3 - z_1}$$

$\Rightarrow \quad (z_2 - z_1)(z_3 - z_1) = (z_3 - z_2)(z_2 - z_3)$

$\Rightarrow \quad z_1^2 + z_2^2 + z_3^2 = z_1z_2 + z_2z_3 + z_3z_1$ ...(1)

Further, since the circumcentre and the centroid of Δ ABC will coincide, we have

$$z_0 = \frac{1}{3}(z_1 + z_2 + z_3)$$

$\Rightarrow \quad (3z_0)^2 = (z_1 + z_2 + z_3)^2$

$\Rightarrow \quad 9z_0^2 = z_1^2 + z_2^2 + z_3^2 + 2(z_1z_2 + z_2z_3 + z_3z_1)$

$\quad = 3(z_1^2 + z_2^2 + z_3^2)$ [$\because$ from (1)]

$\Rightarrow \quad z_1^2 + z_2^2 + z_3^2 = 3z_0^2$, which is true.

Example 22:

Locate the complex numbers

$z = x + iy$ *for which* $\log_{1/2} \dfrac{|z-1| + 4}{3|z-1|-2} > 1.$

Solution:

For logarithm to be positive, we must have

$3 \mid z - 1 \mid - 2 > 0 \Rightarrow 3 \mid z - 1 \mid = 2$

$\Rightarrow \quad \mid z - 1 \mid > \dfrac{2}{3}$ ...(1)

By the law of logarithms, $\dfrac{|z-1| + 4}{3|z-1|-2} < \dfrac{1}{2}$

$\Rightarrow \quad 2 \mid z - 1 \mid + 8 < 3 \mid z - 1 \mid - 2$

$\Rightarrow \quad 10 < \mid z - 1 \mid \Rightarrow \mid z - 1 \mid > 10.$

This does not contradict (1).

Hence, the points z lie in the exterior to the circle with centre (1, 0) and radius 10.

Example 23:

(a) If $(a_1 + ib_1)(a_2 + ib_2) \ldots (a_n + ib_n) = A + iB$, *show that*

(i) $(a_1^2 + b_1^2)(a_2^2 + b_2^2) \ldots (a_n^2 + b_n^2) = A^2 + B^2.$

(ii) $\tan^{-1} \dfrac{b_1}{a_1} + \tan^{-1} \dfrac{b_2}{a_2} + \ldots \tan^{-1} \dfrac{b_n}{a_n} = \tan^{-1} \dfrac{B}{A}.$

(b) If $\tan(\theta + i\phi) = e^{i\alpha}$, *show that*

$$\theta = \left(n + \frac{1}{2}\right)\frac{\pi}{2} \text{ and } \phi = \frac{1}{2}\log\tan\left(\frac{\pi}{4} + \frac{\alpha}{2}\right).$$

Solution:

(a) Let

$$\left.\begin{aligned} a_1 + ib_1 &= r_1(\cos\theta_1 + i\sin\theta_1) \\ a_2 + ib_2 &= r_2(\cos\theta_2 + i\sin\theta_2) \\ &\cdots\cdots \\ a_n + ib_n &= r_n(\cos\theta_n + i\sin\theta_n) \end{aligned}\right\} \quad \ldots(1)$$

so that $r_1^2 = a_1^2 + b_1^2,\ r_2^2 = a_2^2 + b_2^2, \ldots,$
$r_n^2 = a_n^2 + b_n^2$

and $\theta_1 = \tan^{-1}\dfrac{b_1}{a_1},\ \theta_2 = \tan^{-1}\dfrac{b_2}{a_2}, \ldots\ \theta_n = \tan^{-1}\dfrac{b_n}{a_n}.$

Also let $A + iB = R(\cos\theta + i\sin\theta)$ ...(2)

so that $R^2 = A^2 + B^2$ and $\theta = \tan^{-1}\dfrac{B}{A}.$

Thus $(\alpha_1 + ib_1)(\alpha_2 + ib_2)\ldots(\alpha_n + ib_n) = A + iB$ becomes :

$r_1(\cos\theta_1 + i\sin\theta_1)\, r_2(\cos\theta_2 + i\sin\theta_2)\ldots$

$r_n(\cos\theta_n + i\sin\theta_n) = R(\cos\theta + i\sin\theta)$ [Using (1) and (2)]

$\Rightarrow\quad r_1r_2\ldots r_n[\cos(\theta_1 + \theta_2 + \ldots + \theta_n) + i\sin(\theta_1 + \theta_2 + \ldots + \theta_n)] = R(\cos\theta + i\sin\theta)$

$\Rightarrow\quad r_1r_2\ldots r_n = R$ ...(3)

and $\quad \theta_1 + \theta_2 + \ldots + \theta_n = \theta$ ...(4)

(i) $(a_1^2 + b_1^2)(a_2^2 + b_2^2)\ldots(a_n^2 + b_n^2)$
$= r_1^2.r_2^2\ldots r_n^2 = R^2 = A^2 + B^2$

(ii) $\theta_1 + \theta_2 + \ldots + \theta_n = 0$

$\Rightarrow\quad \tan^{-1}\dfrac{b_1}{a_1} + \tan^{-1}\dfrac{b_2}{a_2}, \ldots + \tan^{-1}\dfrac{b_n}{a_n} = \tan^{-1}\dfrac{B}{A}.$

(b) We have:

$\tan(\theta + i\phi) = e^{i\alpha} = \cos\alpha + i\sin\alpha$ ...(1)

$\tan(\theta = i\phi) = \cos\alpha = i\sin\alpha$ ...(2)

Now $\tan 2\theta = \tan\ (\overline{\theta + i\phi} + \overline{\theta - i\phi})$

$$= \frac{\tan(\theta + i\phi) + \tan+(\theta - i\phi)}{1 - \tan(\theta + i\phi)\tan(\theta - i\phi)}$$

$$= \frac{(\cos\alpha + i\sin\alpha) + (\cos\alpha - i\sin\alpha)}{1 - (\cos\alpha + i\sin\alpha)(\cos\alpha - i\sin\alpha)}$$

$$= \frac{2\cos\alpha}{1-(\cos^2\alpha+\sin^2\alpha)} = \frac{2\cos\alpha}{0} = \tan\frac{\pi}{2}$$

$$\therefore \quad 2\theta = n\theta + \frac{\pi}{2}$$

$$\Rightarrow \quad \theta = \left(n+\frac{1}{2}\right)\frac{\pi}{2}$$

And $\tan(2i\phi) = \tan\left(\overline{\theta + i\phi} + \overline{\theta - i\phi}\right)$

$$= \frac{\tan(\theta+i\phi) - \tan+(\theta-i\phi)}{1+\tan(\theta+i\phi)\tan(\theta-i\phi)}$$

$$= \frac{(\cos\alpha+i\sin\alpha)-(\cos\alpha-i\sin\alpha)}{1+(\cos\alpha+i\sin\alpha)(\cos\alpha-i\sin\alpha)}$$

$$\Rightarrow \quad i\,\frac{e^{2\phi}-e^{-2\phi}}{e^{2\phi}+e^{-2\phi}} = \frac{2i\sin\alpha}{1+(\cos^2\alpha+\sin^2\alpha)}$$

$$= \frac{2i\sin\alpha}{1+(\cos^2\alpha+\sin^2\alpha)} = \frac{2i\sin\alpha}{1+1} = i\sin\alpha$$

$$\left[\because \sin 2i\phi = i\left(\frac{e^{2\phi}-e^{-2\phi}}{2}\right), \cos 2u\phi = \frac{e^{2\phi}+e^{-2\phi}}{2}\right]$$

$$\Rightarrow \quad \frac{e^{2\phi}-e^{-2\phi}}{e^{2\phi}+e^{-2\phi}} = \frac{\sin\alpha}{1}$$

By componendo and dividendo,

$$\frac{2e^{2\phi}}{2e^{-2\phi}} = \frac{1+\sin\alpha}{1-\sin\alpha}\ \frac{\left(\cos\frac{\alpha}{2}+\sin\frac{\alpha}{2}\right)^2}{\left(\cos\frac{\alpha}{2}-\sin\frac{\alpha}{2}\right)^2}$$

$$\Rightarrow \quad e^{4\phi} = \left(\frac{1+\tan\alpha/2}{1-\tan\alpha/2}\right)^2$$

$$\Rightarrow \quad e^{4\phi} = \frac{1+\tan\alpha/2}{1-\tan\alpha/2} = \tan\left(\frac{\pi}{4}+\frac{\alpha}{2}\right)$$

$$\Rightarrow \quad 2\phi = \log\tan\left(\frac{\pi}{4}+\frac{\alpha}{2}\right)$$

$$\Rightarrow \quad \phi = \frac{1}{2}\log\tan\left(\frac{\pi}{4}+\frac{\alpha}{2}\right).$$

Example 24:

Determine the region for which

$|z - 1| + |z + 1| \leq 4.$

Solution:

Let $z = x + iy$

in the L.H.S. of given inequality, we have

$$|z - 1| + |z + 1| = |x + iy - 1| + |x + iy + 1|$$

$$= |(x - 1) + iy| + |(x + 1) + iy|$$

$$= \sqrt{(x-1)^2 + y^2} + \sqrt{(x+1)^2 + y^2}$$

Thus, $\sqrt{(x-1)^2 + y^2} + \sqrt{(x+1)^2 + y^2} \leq 4$

$$\Rightarrow \quad \sqrt{(x+1)^2 + y^2} \leq 4 - \sqrt{(x-1)^2 + y^2}$$

Squaring both sides, we have

$$(x + 1)^2 + y^2 \leq [4 - \sqrt{(x-1)^2 + y^2}]^2$$

$$\Rightarrow \quad (x + 1)^2 + y^2 \leq 16 + (x - 1)^2 + y^2 - 8\sqrt{(x-1)^2 + y^2}$$

$$\Rightarrow \quad 8\sqrt{(x-1)^2 + y^2} \leq 16 + (x - 1)^2 - (x + 1)^2$$

$$\Rightarrow \quad 8\sqrt{(x-1)^2 + y^2} \leq 16 - 4x$$

$$\Rightarrow \quad 2\sqrt{(x-1)^2 + y^2} \leq 4 - x$$

Squaring both sides, we have

$$4[x - 1)^2 + y^2] \leq (4 - x)^2$$

$$\Rightarrow \quad 4x^2 - 8x + 4 + 4y^2 \leq 16 - 8x + x^2$$

$$\Rightarrow \quad 3x^2 + 4y^2 \leq 12$$

$$\Rightarrow \quad \frac{x^2}{4} + \frac{y^2}{3} \leq 1.$$

The equation $\frac{x^2}{4} + \frac{y^2}{3} = 1$ represents an ellipse.

Hence, the relation $\frac{x^2}{4} + \frac{y^2}{3} \leq 1$ represents the region, which is interior and boundary of the ellipse

$$\frac{x^2}{4} + \frac{y^2}{3} = 1.$$

Example 25:

If $z = 2 + t + i\sqrt{3 - t^2}$, *where t is real and* $t^2 < 3$, *show that the modulus of* $\frac{z+1}{z-1}$ *is independent of t. Also show that the locus of the points z for different values of t is a circle and find its centre and radius.*

Solution:

(i) We have : $z = 2 + t + i\sqrt{3 - t^2}$

$$\therefore \quad \frac{z+1}{z-1} = \frac{3 + t + i\sqrt{3 - t^2}}{1 + t + i\sqrt{3 - t^2}}$$

$$\therefore \quad \left|\frac{z+1}{z-1}\right|^2 = \frac{(3+t)^2 + (3 - t^2)}{(1+t)^2 + (3 - t^2)} = \frac{12 + 6t}{4 + 2t} = 3$$

$$\Rightarrow \quad \left|\frac{z+1}{z-1}\right| = \sqrt{3}, \text{ which is independent of t.}$$

(ii) Let $z = x + iy$.

Then $z = 2 + t + i\sqrt{3 - t^2}$

$\Rightarrow \quad x + iy = 2 + t + i\sqrt{3 - t^2}$

$\Rightarrow \quad x = 2 + t$ and $y = \sqrt{3 - t^2}$

$\Rightarrow \quad x - 2 = 1$ and $y = \sqrt{3 - t^2}$

$\Rightarrow \quad (x - 2)^2 = t^2$ and $y^2 = 3 - t^2$

$\Rightarrow \quad (x - 2)^2 = 3 - y^2 \Rightarrow (x - 2)^2 + y^2 = (\sqrt{3})^2.$

Hence, locus of points z is a circle with centre $2 + i0$ (*i.e.*, (2, 0)) and radius $\sqrt{3}$.

Example 26:

Suppose the points $z_1, z_2, \ldots, z_n$ *all lie on one side of a line drawn through the origin of the complex plane. Prove that the same of the points* $\frac{1}{z_1}, \frac{1}{z_2}, \ldots, \frac{1}{z_n}$. *Moreover, show that*

$$z_1 + z_2 + \ldots + z_n \neq 0, \quad \frac{1}{z_1} + \frac{1}{z_2} + \ldots + \frac{1}{z_n} \neq 0.$$

Solution:

The equation of any line passing through the origin is $\bar{a}z + a\bar{z} = 0$.

Since $z_1, z_2, \ldots, z_n$ lie in the half-plane $\bar{a}z + a\bar{z} > 0$.

$\therefore \bar{a}z_i + a\bar{z}_i > 0$ for $i = 1, 2, \ldots, n$

$$\Rightarrow \quad \bar{a}\left(\frac{z_i\bar{z}_i}{\bar{z}_i}\right) + a\left(\frac{\bar{z}_i z_i}{z_i}\right) > 0$$

$$\Rightarrow \quad \bar{a}\left(\frac{|z_i|^2}{\bar{z}_i}\right) + a\left(\frac{|z_i|^2}{z_i}\right) > 0$$

$$\Rightarrow \quad |z_i|^2\left(\frac{\bar{a}}{\bar{z}_i} + \frac{a}{z_i}\right) > 0$$

$$\Rightarrow \quad a\left(\frac{1}{z_i}\right) + \bar{a}\left(\frac{1}{\bar{z}_i}\right) > 0 \text{ for } i = 1, 2, \ldots n \qquad [\because |z_i|^2 > 0]$$

$$\Rightarrow \quad \frac{1}{z_1}, \frac{1}{z_2}, \ldots, \frac{1}{z_n} \text{ lie on one side of the line } \bar{a}z + a\bar{z} = 0.$$

Also $\bar{a}z_i + a\bar{z}_i > 0$ for $i = 1, 2, \ldots, n$

$$\Rightarrow \quad \bar{a}\left(\sum_{i=1}^{n} z_i\right) + a\left(\sum_{i=1}^{n} \bar{z}_i\right) > 0$$

$$\Rightarrow \quad \bar{a}\left(\sum_{i=1}^{n} z_i\right) + a\overline{\left(\sum_{i=1}^{n} z_i\right)} > 0$$

$$\Rightarrow \quad \sum_{i=1}^{n} z_i \neq 0.$$

Similarly $\sum_{i=1}^{n} \frac{1}{z_i} \neq 0$.

Example 27:

If $x_1, x_2, \ldots, x_n$ are the roots of the equation
$x^n + p_1x^{n-1} + p_2x^{n-2} + \ldots + p_{n-1}x + p_n = 0$,
where co-efficients $p_1, p_2, \ldots, p_n$ are real, show that

$$(1 + x_1^2)(1 + x_2^2)(1 + x_3^2) \ldots (1 + x_n^2)$$
$$= (1 - p_2 + p_4 \ldots)^2 + (p_1 - p_3 + p_5 \ldots)^2.$$

Solution:

Since $x_1, x_2, x_3, \ldots, x_n$ are the roots of the given equation, we have the identity

$$x^n + p_1x^{n-1} + p_2x^{n-2} + \ldots + p_{n-1}x + p_n$$
$$= (x - x_1)(x - x_2)(x - x_3) \ldots (x - x_n)$$

Replacing x by i and –i respectively, we get

$$i^n + p_1i^{n-1} + p_2i^{n-2} + \ldots + p_{n-1}i + p_n$$
$$= (i - x_1)(i - x_2)(i - x_3) \ldots (i - x_n) \qquad \ldots(1)$$

and $(-i)^n + p_1(-i)^{n-1} + p_2(-i)^{n-2} + \ldots + p_{n-1}(-i) + p_n$

$$= (-i - x_1)(-i - x_2)(-i - x_3) \ldots (-i - x_n) \qquad \ldots(2)$$

Now consider L.H.S. of (1). It is

$$i^n\left[1 + p_1.\frac{1}{i} + p_2.\frac{1}{i^2} + p_3.\frac{1}{i^3} + p_4.\frac{1}{i^4} + \ldots\right]$$
$$= i^n[1 - p_1i - p_2 + p_3i + p_4 \ldots]$$
$$= i^n[(1 - p_2 + p_4 \ldots) = i(p_1 - p_3 + \ldots)] \qquad \ldots(3)$$

Similarly R.H.S. of (2) becomes

$$(-i)^n\left[1 + p_1\left(-\frac{1}{i}\right) + p_2\left(-\frac{1}{i}\right)^2 + p_3\left(-\frac{1}{i}\right)^3 + p_4\left(-\frac{1}{i}\right)^4 + \ldots\right]$$
$$= (-i)^n[1 + p_1i - p_2 - p_3i + p_4 + \ldots]$$
$$= (-1)^n[(1 - p_2 + p_4 \ldots) + i(p_1 - p_3 + \ldots)] \qquad \ldots(4)$$

Now multiplying (1) and (2) after replacing their L.H. sides by (3) and (4) respectively, we get

$$i^n[(1 - p_2 + p_4 \ldots) - i(p_1 - p_2 + \ldots)]$$
$$\times (-i)^n[(1 - p_2 + p_4 \ldots) + i(p_1 - p_2 + \ldots)]$$
$$= (i - x_1)(i - x_2)(i - x_3) \ldots (i - x_n)$$
$$\times (-i - x_1)(-i - x_2)(-i - x_3) \ldots (-i - x_n)$$
$$\Rightarrow (-i^2)^n[(1 - p_2 + p_4 \ldots)^2 + (p_1 - p_3 + \ldots)^2]$$
$$= (-1)^n[(i - x_1)(i + x_1).(i - x_2)(i + x_2) \ldots (i - x_n)(i + x_n)]$$

Hence $(1 - p_2 + p_4 \ldots)^2 + (p_1 - p_3 + \ldots)^2$

$$= (-1)^n[(-1 - x_1^2).(-1 - x_2^2) \ldots (-1 - x_n^2)]$$

$= (-1)^{2n}[(1 + x_1^2)(1 + x_2^2) \dots (1 + x_n^2)]$
$= (1 + x_1^2)(1 + x_2^2) \dots (1 + x_n)^2.$

Example 28:

Prove that the complex numbers z_1, z_2 and the origin form an equilateral triangle only, if

$$z_1^2 + z_2^2 - z_1 z_2 = 0.$$

Solution:

B(Z_2)

A(Z_1)

O(0)

Since Δ OAB is equilateral, we have

OA = OB = AB

$$\Rightarrow \quad \frac{OB}{OA} = \frac{AB}{OB}$$

$$\Rightarrow \quad \left|\frac{z_2 - 0}{z_1 - 0}\right| = \left|\frac{z_2 - z_1}{z_2 - 0}\right|$$

$$\Rightarrow \quad \frac{z_2}{z_1} = \frac{z_2 - z_1}{z_2} \Rightarrow z_2^2 = z_1 z_2 - z_1^2$$

$$\Rightarrow \quad z_1^2 + z_2^2 - z_1 z_2 = 0, \text{ which is true.}$$

Example 29:

Show that the triangles whose vertices are z_1, z_2, z_3 and a, b, c are similar if

$$\begin{vmatrix} z_1 & a & 1 \\ z_2 & b & 1 \\ z_3 & c & 1 \end{vmatrix} = 0.$$

Solution:

Let z_1, z_2, z_3 be given by A, B, C respectively and a, b, c be given by D, E, F respectively.

Since Δ ABC and Δ DEF are similar,

$$\therefore \frac{AB}{DE} = \frac{BC}{EF} \text{ and } \angle ABC = \angle DEF = \alpha \text{ (say)}$$

$$\text{Thus } \alpha = \arg\left(\frac{z_1 - z_2}{z_3 - z_2}\right) = \arg\left(\frac{a - b}{c - b}\right)$$

$$\Rightarrow \quad \frac{z_1 - z_2}{z_3 - z_2} = \frac{AB}{BC} (\cos\alpha + i \sin\alpha) \qquad \dots(1)$$

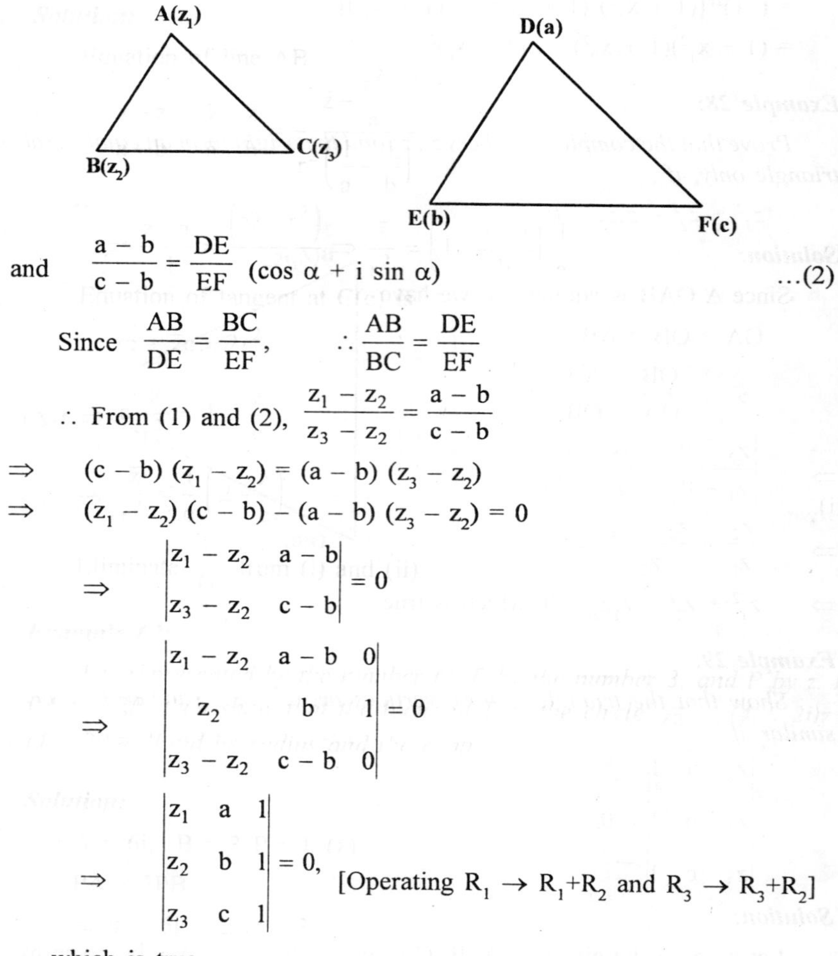

and $$\frac{a - b}{c - b} = \frac{DE}{EF}\ (\cos\alpha + i\sin\alpha) \qquad \ldots(2)$$

Since $\dfrac{AB}{DE} = \dfrac{BC}{EF}$, $\quad \therefore \dfrac{AB}{BC} = \dfrac{DE}{EF}$

$\therefore$ From (1) and (2), $\dfrac{z_1 - z_2}{z_3 - z_2} = \dfrac{a - b}{c - b}$

$\Rightarrow \quad (c - b)(z_1 - z_2) = (a - b)(z_3 - z_2)$

$\Rightarrow \quad (z_1 - z_2)(c - b) - (a - b)(z_3 - z_2) = 0$

$$\Rightarrow \quad \begin{vmatrix} z_1 - z_2 & a - b \\ z_3 - z_2 & c - b \end{vmatrix} = 0$$

$$\Rightarrow \quad \begin{vmatrix} z_1 - z_2 & a - b & 0 \\ z_2 & b & 1 \\ z_3 - z_2 & c - b & 0 \end{vmatrix} = 0$$

$$\Rightarrow \quad \begin{vmatrix} z_1 & a & 1 \\ z_2 & b & 1 \\ z_3 & c & 1 \end{vmatrix} = 0, \quad [\text{Operating } R_1 \to R_1 + R_2 \text{ and } R_3 \to R_3 + R_2]$$

which is true.

Example 30:

Find the tenth roots of unity and show that the product of any two of them is again one of the tenth roots of unity.

Solution:

Since $\quad 1 = \cos 0 + i \sin 0$

$\qquad\quad = \cos 2n\pi + i \sin 2n\pi;\ n \in I$

$\therefore \quad (1)^{1/10} = (\cos 2n\pi + i \sin 2n\pi)^{1/10}$

$= \cos \frac{n\pi}{5} + i \sin \frac{n\pi}{5}$; n = 1, 2, ..., 9

n = 0; $\cos 0 + i \sin 0 = 1 = r_1$

n = 1; $\cos \frac{\pi}{5} + \sin \frac{\pi}{5} = r_2$

n = 2; $\cos \frac{2\pi}{5} + i \sin \frac{2\pi}{5} = r_3$

....

n = 9; $\cos \frac{9\pi}{5} + i \sin \frac{9\pi}{5} = r_{10}$

Thus, the tenth roots of unity are obtained by given n = 0, 1, 2, ..., 9.

It can be verified that the product of any two of the above is equal to one of them.

For example, $r_3 . r_8 = \left(\cos \frac{3\pi}{5} + i \sin \frac{3\pi}{5}\right)\left(\cos \frac{8\pi}{5} + i \sin \frac{8\pi}{5}\right)$

$$= \cos \frac{11\pi}{5} + i \sin \frac{11\pi}{5}$$

$$= \cos\left(2\pi + \frac{\pi}{5}\right) + i \sin\left(2\pi + \frac{\pi}{5}\right)$$

$$= \cos \frac{\pi}{5} + i \sin \frac{\pi}{5} = r_2.$$ And so on.

Example 30(a):

If 1, ω, ω^2 are three cube roots of unity and $x = a + b$, $y = a\omega + b\omega^2$, $z = a\omega^2 + b\omega$, then prove that $x^2 + y^2 + z^2 = 6ab$.

Solution:

We have $x = a + b$, $y = a\omega + b\omega^2$; $z = a\omega^2 + b\omega$.

L.H.S. $= x^2 + y^2 + z^2$

$= (a + b)^2 + (a\omega + b\omega^2)^2 + (a\omega^2 + b\omega)^2$

$= a^2 + b^2 + 2ab + a^2\omega^2 + b^2\omega^4 + 2ab\omega^3 + a^2\omega^4 + b^2\omega^2 + 2ab\omega^3$

$= a^2 + b^2 + 2ab + a^2\omega^2 + b^2\omega + 2ab\ (1) + a^2\omega + b^2\omega^2 + 2ab\ (1)$

$[\because \omega^2 = 1$ and $\omega^4 = \omega^3.\omega = 1.\ \omega = \omega]$

$= a^2(1 + \omega + \omega^2) + b^2(1 + \omega + \omega^2) + 6ab$

$= a^2(0)\ b^2(0) + 6ab \quad [\because 1 + \omega + \omega^2 = 0]$

$= 6ab =$ R.H.S.

Example 31:

Determine the maximum distance from the origin of co-ordinates to the points z for which $\left|z + \frac{1}{z}\right| = 1.$

Solution:

We have $\left|z + \frac{1}{z}\right| = 1$

$\Rightarrow \quad z^2 + 1 \,| = | \, z \,|$

Putting $z = r(\cos\theta + i\sin\theta)$, we get

$$|\, r^2 (\cos\theta + i\sin\theta)^2 + r \,| = r$$

$$\Rightarrow \quad |\, r^2 (\cos 2\theta + i\sin 2\theta) + 1 \,| = r$$

$$\Rightarrow \quad |\, (r^2 \cos 2\theta + 1) + i\, r^2 \sin 2\theta \,| = r$$

$$\Rightarrow \quad \sqrt{(r^2\cos 2\theta + 1)^2 + (r^2 \sin 2\theta)^2} = r$$

$$\Rightarrow \quad (r^2 \cos 2\theta + 1)^2 + r^4 \sin 2\theta = r^2$$

$$\Rightarrow \quad r^4\cos^2 2\theta + 2r^2\cos 2\theta + 1 + r^4\sin^2 2\theta = r^2$$

$$\Rightarrow \quad r^4 + (2\cos 2\theta - 1)r^2 + 1 = 0$$

$$\therefore\ r^2 = \frac{(1-2\cos 2\theta) \pm \sqrt{(-2\cos 2\theta + 1)^2 - 4}}{2}$$

r^2 is maximum when $-2\cos 2\theta + 1$ is maximum

i.e., when $\cos 2\theta = -1$

$$\therefore\ \text{Max. value of } r^2 = \frac{3+\sqrt{5}}{2} = \frac{6+2\sqrt{5}}{4}.$$

Hence max. value of $r = \frac{1}{2}\sqrt{6+2\sqrt{5}}$

$$= \frac{1}{2}\sqrt{(\sqrt{5}+1)^2} = \frac{1}{2}(\sqrt{5}+1).$$

Example 32:

Find the greatest and the least values of the moduli of complex numbers z satisfying the equation $\left|z - \frac{4}{z}\right| = 2$. *Find also the corresponding complex numbers.*

Solution:

We have $\left| |z| - \left|\frac{4}{z}\right| \right| \le \left| z - \frac{4}{z} \right| = 2$

$\Rightarrow \quad -2 \le |z| - \frac{4}{|z|} \le 2$

$\Rightarrow \quad |z|^2 + 2|z| - 4 \ge 0$ or $|z|^2 - 4 \le 0$

$\Rightarrow \quad (|z| + 1)^2 - 5 \ge 0$ or $(|z| - 1)^2 \le 5$

$\Rightarrow \quad (|z| + 1 + \sqrt{5})\ (|z| + 1 - \sqrt{5}) \ge 0$

or $\quad (|z| - 1 + \sqrt{5})\ (|z| - 1 - \sqrt{5} \le 0$

$\Rightarrow \quad |z| \le -\sqrt{5} - 1$ or $|z| \ge \sqrt{5} - 1$

or $\quad \sqrt{5} - 1 \le |z| \le \sqrt{5} + 1$

$\Rightarrow$ Greatest value of $|z| = \sqrt{5} + 1$ and least value of $|z| = \sqrt{5} - 1$.

$|z| = \sqrt{5} + 1 \Rightarrow \frac{4}{|z|} = \sqrt{5} - 1$

Now, $\quad \frac{4}{z} = \frac{4\bar{z}}{|z|^2} \Rightarrow \frac{4}{z}$ lying in the direction of $\bar{z}$

$\left| z - \frac{4}{z} \right| = PR = 2$ (given)

We have $OP = \sqrt{5} + 1$ and $OR = \sqrt{5} - 1$

$$\Rightarrow \quad \cos 2\theta = \frac{OP^2 + OR^2 - PR^2}{2OP.OR}$$

$$= \frac{\left(\sqrt{5}+1\right)^2 + \left(\sqrt{5}-1\right)^2}{2(5-1)} = 1$$

$\Rightarrow \quad 2\theta = 0, 2\pi \Rightarrow Z = \left(\sqrt{5}-1\right)$

Similarly for $|z| = \sqrt{5} - 1$

we get $z = \left(\sqrt{5}-1\right)$

Alternative Solution: In ΔOPQ

$$\cos 2\theta = \frac{OP^2 - OR^2 - PR^2}{2OP.OR} = \frac{|z| - \left|\frac{4}{z}\right|^2 - \left|z - \frac{4}{z}\right|^2}{2|z|.\left|\frac{4}{z}\right|}$$

$$= \frac{r^2 + \frac{16}{r^2} - 2^2}{8} \text{ (Let } |z| = r > 0) = \frac{\left(r + \frac{4}{r}\right) - 12}{8}$$

Since $-1 \leq \cos 2\theta \leq 1$ ...(1)

$\Rightarrow \quad -1 \leq = \dfrac{\left(r+\frac{4}{r}\right)^2 - 12}{8}$

$\Rightarrow \quad -8 \leq \left(r+\frac{4}{r}\right)^2 - 12 \leq 8$

$\Rightarrow \quad \left(r+\frac{4}{r}\right)^2 \geq 4$, which is always true.

And $\quad \left(r+\frac{4}{r}\right)^2 \leq 20 \Rightarrow r + \frac{4}{r} \leq 2\sqrt{5}$ (as $r > 0$)

$\Rightarrow \quad r^2 - 2 2\sqrt{5}\ r + 4 £ 0$

$\Rightarrow \quad \left(r - \sqrt{5} - 1\right)\left(r - \left(\sqrt{5}+1\right)\right) \leq 0$

$\Rightarrow \quad \sqrt{5} - 1 \leq r \leq \sqrt{5} + 1$

Hence max. $r = \sqrt{5}+1$ and min. $r = \sqrt{5}-1$

and it occurs at the right end of the inequality (1).

Hence $\cos 2\theta = 1 \Rightarrow 2\theta = 0$ or 2π

$\Rightarrow \quad \theta = 0$ or π

$\Rightarrow$ for $\quad r = \sqrt{5} + 1$, $z = \left(\sqrt{5}+1\right)$

and for $\quad r = \sqrt{5} - 1$, $z = \pm\left(\sqrt{5}-1\right)$.

Example 33:

Consider a quadratic equation $az^2 + bz + c = 0$ where a, b, c, are complex numbers. Find the condition that the equation has

(i) one purely imaginary root

(ii) one purely real root

(iii) two purely imaginary roots

(iv) two purely real root

Solution:

(i) Let z_1 (purely imaginary) be a root of the give equation.

$\Rightarrow z_1 = -\ \bar{z}_1$

and $az_1^2 + bz_1 + c = 0$ ...(1)

$\Rightarrow \overline{az_1^2 + bz_1 + c} = \bar{0}$

$\Rightarrow \overline{az_1^2} + \overline{bz_1} + \bar{c} = 0$

$\Rightarrow \bar{a}z_1^2 - \bar{b}z_1 + \bar{c} = 0$ (as $\bar{z}_1 = -z_1$) ...(2)

Now (1) and (2) must have one common root.

$$\Rightarrow \frac{z_1^2}{\left(b\bar{c} + c\bar{b}\right)} = \frac{z_1}{\left(c\bar{a} - a\bar{c}\right)} = \frac{1}{-\left(a\bar{b} + \bar{a}b\right)}$$

$$\Rightarrow z_1^2 = -\frac{\left(b\bar{c} + c\bar{b}\right)}{\left(a\bar{b} + \bar{a}b\right)} = \frac{\left(c\bar{a} - a\bar{c}\right)^2}{\left(a\bar{b} + \bar{a}b\right)^2}$$

$$\Rightarrow \left(b\bar{c} + c\bar{b}\right)\left(a\bar{b} + \bar{a}b\right) + \left(c\bar{a} - a\bar{c}\right)^2 = 0$$

(ii) Let z_1 (purely real, be a root of the given equation

$\Rightarrow \quad z_1 = \bar{z}_1$

and $az_1^2\ bz_1 + c = 0$ ...(1)

$\Rightarrow \overline{az_1^2 + bz_1 + c} = \bar{0}$

$\Rightarrow \overline{az_1^2} + \overline{bz_1} + \bar{c} = 0$

$\Rightarrow \bar{a}z_1^2 - \bar{b}z_1 + \bar{c} = 0$ (as $\bar{z}_1 = -z_1$) ...(2)

Now (1) and (2) must have one common root.

$$\Rightarrow \frac{z_1^2}{\left(b\bar{c} - \bar{b}c\right)} = \frac{z_1}{\left(a\bar{c} - \bar{a}c\right)} = \frac{1}{\left(a\bar{b} - \bar{a}b\right)}$$

Required condition is

$$\Rightarrow \left(b\bar{c} - c\bar{b}\right)\left(a\bar{b} - \bar{a}b\right) + \left(c\bar{a} - a\bar{c}\right)^2 = 0$$

(iii) Let z_1 and z_2 be two purely imaginary roots then

$\bar{z}_1 = -z_1, \bar{z}_2 = -z_2$

Now $az^2 + bz + c = 0$ ...(1)

$$\Rightarrow \qquad az^2 + bz + c = \overline{0}$$

$$\Rightarrow \qquad \overline{az}^2 + \bar{b}\bar{z} + \bar{c} = 0$$

$$\Rightarrow \qquad \bar{a}z^2 - \bar{b}z + \bar{c} = 0 \qquad ...(2)$$

Equation (1) and (2) must be identical as their root are same

$$\Rightarrow \qquad \frac{a}{a} = -\frac{b}{b} = \frac{c}{c}$$

(iv) Let z_1 and z_2 be two purely real roots then

$$\bar{z}_1 = z_1, \bar{z}_2 = z_2$$

In this case $az^2 + bz + c = 0$ ...(1)

$$\Rightarrow \qquad \overline{az}^2 + \bar{b}\bar{z} + \bar{c} = 0$$

$$\Rightarrow \qquad \bar{a}z^2 + \bar{b}z + \bar{c} = 0 \qquad ...(2)$$

Equation (1) and (2) must be identical, as their root are same

$$\Rightarrow \qquad \frac{a}{\bar{a}} = \frac{b}{\bar{b}} = \frac{c}{\bar{c}}.$$

Example 34:

Let z_1, z_2, z_3 three distinct complex numbers satisfying $|z_1 - 1| = |z_2 - 1| = |z_3 - 1|$. Let A, B, and C be the points represented in the Argand plane corresponding to z_1, z_2 and z_3 respectively. Prove that $z_1 + z_2 + z_3 = 3$ if and only of Δ ABC is an equilateral triangle.

Solution:

$$|z_1 - 1| = |z_2 - 1| = |z_3 - 1|$$

$\Rightarrow$ The point corresponding to 1 (say P) is equidistant from the points A, B and C.

$\Rightarrow$ P is the circumcentre of the ΔABC

Now if $z_1 + z_2 + z_3 = 3$ then the point corresponding to centroid of the ΔABC is $\frac{z_1 + z_2 + z_3}{3} = 1 \Rightarrow$ circumcentre and centroid coincide.

$\Rightarrow$ ΔABC is equilateral

Conversely if ΔABC is equilateral, then centroid is the same as the circumcentre *i.e.*, P. Hence centroid $\frac{z_1 + z_2 + z_3}{3} = 1$

$\Rightarrow z_1 + z_2 + z_3 = 3.$

Example 35:

Consider two pairs of non-zero conjugate complex numbers (z_1, z_2) and (z_3, z_4). Find the value of arg

$$\left(\frac{z_1}{z_3}\right) + \arg\left(\frac{z_2}{z_4}\right).$$

Solution:

$$\arg\left(\frac{z_1}{z_3}\right) + \arg\left(\frac{z_2}{z_4}\right) = \arg\left(\frac{z_1 z_2}{z_3 z_4}\right)$$

$$= \arg\left(\frac{|z_1|^2}{|z_3|^2}\right) \quad (\text{as } z_2 = \bar{z}_1 \text{ and } z_4 = \bar{z}_3$$

= 0 (as argument of a positive real number is zero).

Example 36:

If $|z_1| = |z_2| = |z_3| = 1$, prove that $|z_1 + z_2 + z_3| = = \left|\frac{1}{z_1} + \frac{1}{z_2} + \frac{1}{z_3}\right|$.

Solution:

We know that $|z| = |\bar{z}|$

$$\Rightarrow \qquad |z_1 + z_2 + z_3| = |\bar{z}_1 + \bar{z}_2 + \bar{z}_3|$$

$$= \frac{\bar{z}_1 . z_1}{z_1} + \frac{\bar{z}_2 . z_2}{z_2} + \frac{\bar{z}_3 . z_3}{z_3}$$

$$= \left|\frac{|z_1|^2}{z_1} + \frac{|z_2|^2}{z_2} + \frac{|z_3|^2}{z_3}\right|$$

$$= \left|\frac{1}{z_1} + \frac{1}{z_2} + \frac{1}{z_3}\right| \qquad (\because |z_1)^2 = |z_2|^2 = |z_3|)^2 = 1)$$

Example 37:

If z_1 and z_2 are two complex numbers, then prove that
$|z_1 - z_2|^2 \leq (1 + k)\, |z_1|^2 + (1 + k^{-1})\, |z_2|^2 \;\; \forall k \in R.$

Solution:

$$|(kz_1 + z_2)|^2 \geq 0 \Rightarrow (kz_1 + z_2)\left(k\bar{z}_1 + \bar{z}_2\right) \geq 0$$

$$\Rightarrow \left(z_1 + \frac{z_2}{k}\right)\left(k\bar{z}_1 + \bar{z}_2\right) \geq 0 \text{ as } k > 0$$

$$\Rightarrow k\,|z_1|^2 + \frac{1}{k}\,|z_2|^2 + z_1\bar{z}_2 + \bar{z}_1 z_2 \geq 0$$

$$\Rightarrow (1 + k)\,|z_1)^2 + (1 + k^{-1})\,|z_2)^2 \geq |z_1|^2 + |z_2|^2 - \bar{z}_1 z_2 - \bar{z}_1 z_2$$

$$\Rightarrow (1 + k)\,|z_1|^2 + (1 + k^{-1})\,|z_2|^2 \geq |z_1 - z_2|^2.$$

Example 38:

Solve the equation $z^7 + 1 = 0$ and hence deduce that

(i) $\cos\frac{\pi}{7}\cos\frac{3\pi}{7}\cos\frac{5\pi}{7} = -\frac{1}{8}$

(ii) $\cos\frac{\pi}{14}\cos\frac{3\pi}{14}\cos\frac{5\pi}{14} = \frac{\sqrt{7}}{8}$

(iii) $\sin\frac{\pi}{14}\sin\frac{3\pi}{14}\sin\frac{5\pi}{14} = \frac{1}{8}$

(iv) $\tan^2\frac{\pi}{14} + \tan^2\frac{3\pi}{14} + \tan^2\frac{5\pi}{14} = 5$

Solution:

$$z^7 + 1 = 0$$

The roots are $-1, \alpha, \alpha^3, \alpha^5, \bar{\alpha}, \bar{\alpha}^3, \bar{\alpha}^5$

Where $a = \cos\ \cos\frac{\pi}{7} + i\sin\frac{3\pi}{7}$

$$\alpha^3 = \cos\frac{3\pi}{7} + i\sin\frac{3\pi}{7}$$

$$\alpha^5 = \cos\frac{5\pi}{7} + i\sin\frac{5\pi}{7}$$

Now $z^7 + 1 = (z + 1)(z - a)(z - \bar{\alpha})(z - a^3)(z - a^5)(z - \bar{\alpha}^5)$

$$\Rightarrow \quad \frac{z^7+1}{z+1} = \left\{z^2 - (\alpha+\alpha)z + \alpha\bar{\alpha}\right\}\ldots\left\{z^2 - \left(\alpha^5 + \bar{\alpha}^5\right) + \alpha^5\bar{\alpha}^5\right\}$$

$$\Rightarrow \quad (z^6 - z^5 + z^4 - z^3 + z^2 - z + 1)$$

$$=\left(z^2-2z\cos\frac{\pi}{7}+1\right)\left(z^2-2z\cos\frac{3\pi}{7}+1\right)\left(z^2-2z\cos\frac{5\pi}{7}+1\right) \quad ...(1)$$

[since $a+\overline{\alpha}=2\cos\frac{\pi}{7}$, $\alpha\overline{\alpha}=1$]

$$\Rightarrow \frac{z^6+1}{z^3}-\frac{z^5+z}{z^3}+\frac{z^4+z^2}{z^3}-\frac{z^3}{z^3}$$

$$=\left(\frac{z^2-2z\cos\frac{\pi}{7}+1}{z}\right)\left(\frac{z^2-2z\cos\frac{3\pi}{7}+1}{z}\right)\left(\frac{z^2-2z\cos\frac{5\pi}{7}+1}{z}\right)$$

$$\Rightarrow \left(z_3+\frac{1}{z^3}\right)-\left(z^2+\frac{1}{z^2}\right)+\left(z+\frac{1}{z}\right)-1$$

$$=\left\{\left(z+\frac{1}{z}\right)-2\cos\frac{\pi}{7}\right\}\left\{\left(z+\frac{1}{z}\right)-2\cos\frac{3\pi}{7}\right\}\left\{\left(z+\frac{1}{z}\right)-2\cos\frac{5\pi}{7}\right\}$$

or, $\left(z+\frac{1}{z}\right)^3-3\left(z+\frac{1}{z}\right)-\left(z+\frac{1}{z}\right)^2+2+\left(z+\frac{1}{z}\right)-1$

$$=\left(z+\frac{1}{z}-2\cos\frac{\pi}{7}\right)\left(z+\frac{1}{z}-2\cos\frac{3\pi}{7}\right)\left(z+\frac{1}{z}-2\cos\frac{5\pi}{7}\right)$$

Let $z+\frac{1}{z}=2x$

We get, $8x^3-6x-4x^2+2+2x-1$

$$=8\left(x-\cos\frac{\pi}{7}\right)\left(x-\cos\frac{3\pi}{7}\right)\left(x-\cos\frac{5\pi}{7}\right)$$

or, $8x^3-4x^2-4x+1=8\left(x-\cos\frac{\pi}{7}\right)\left(x-\cos\frac{3\pi}{7}\right)\left(x-\cos\frac{5\pi}{7}\right)=0$ ...(2)

$\therefore$ $\cos\frac{\pi}{7}$, $\cos\frac{3\pi}{7}$ and $\cos\frac{5\pi}{7}$ are the roots of the equation

$$8x^3-4x^2-4x+1=0$$

$\therefore$ Product of roots $= -\ -\frac{1}{8}$

$$\therefore\ \cos\frac{\pi}{7}.\cos\frac{3\pi}{7}.\cos\frac{5\pi}{7}=-\frac{1}{8}.$$

(ii) Putting $x=-1$ in (2), we get

$-8-4+4+1=-8$ $\left(1+\cos\frac{\pi}{7}\right)\left(1+\cos\frac{\pi}{7}\right)\left(1+\cos\frac{3\pi}{7}\right)\left(1+\cos\frac{5\pi}{7}\right)$

or, $7 = 8^2 \cos^2 \frac{\pi}{14}\cos^2\frac{3\pi}{14}\cos^2\frac{5\pi}{14}$

or $\cos\frac{\pi}{14}\cos\frac{3\pi}{14}\cos\frac{5\pi}{14}=\pm\frac{\sqrt{7}}{8}$

since $0 < \frac{\pi}{14},\frac{3\pi}{14},\frac{5\pi}{14}<\frac{\pi}{2}$, we reject the –ve sign

$\therefore\ \cos\frac{\pi}{14}\cos\frac{3\pi}{14}\cos\frac{5\pi}{14}=\frac{\sqrt{7}}{8}$.

(iii) Putting x = 1 in (2), we get

$8-4-4+1=8\left(1-\cos\frac{\pi}{7}\right)\left(1-\cos\frac{3\pi}{7}\right)\left(1-\cos\frac{5\pi}{7}\right)$

or, $1=8^{2\sin^2}\frac{\pi}{14}\sin^2\frac{3\pi}{14}\sin^2\frac{5\pi}{14}$

$\sin\frac{\pi}{14}\sin\frac{3\pi}{14}\sin\frac{5\pi}{14}=\pm\frac{\sqrt{1}}{8}=\pm\frac{1}{8}$

since $0<\frac{\pi}{14},\frac{3\pi}{14},\frac{5\pi}{14}<\frac{\pi}{2}$, we reject the –ve sign

$\therefore\ \sin\frac{\pi}{14}\sin\frac{3\pi}{14}\sin\frac{5\pi}{14}=\pm\frac{\sqrt{1}}{8}=\frac{1}{8}$

(iv) $\tan\frac{\pi}{14}\tan\frac{3\pi}{14}\tan\frac{5\pi}{14}=\frac{1/8}{\sqrt{7}/8}=\frac{1}{\sqrt{7}}$

We have (1),

$$\frac{z^7+1}{z+1}=\left(z^2-2z\cos\frac{\pi}{7}+1\right)\left(z^2-2z\cos\frac{3\pi}{7}+1\right)\left(z^2-2z\cos\frac{5\pi}{7}+1\right)$$

Let $z=\frac{1+y}{1-y}$

$\therefore\ z^2-2z\cos$

$$\cos\frac{\pi}{7}+1=\left(\frac{1+y}{1-y}\right)^2-2\left(\frac{1+y}{1-y}\right)\left(\frac{1-\tan^2\frac{\pi}{14}}{1+\tan^2\frac{\pi}{14}}\right)+1 \quad ...(3)$$

$$=\frac{2\left(1+y^2\right)\left(1+\tan^2\frac{\pi}{14}\right)-2\left(1-y^2\right)\left(1-\tan^2\frac{\pi}{14}\right)}{(1-y)^2\left(1-\tan^2\frac{\pi}{14}\right)}$$

$$=\frac{2\left(2y^2+2\tan^2\frac{\pi}{14}\right)}{(1-y)^2\left(1+\dfrac{\sin^2\frac{\pi}{14}}{\cos^2\frac{\pi}{14}}\right)}=\frac{2^2\cos^2\frac{\pi}{14}}{(1-y)^2}\left(y^2+\tan^2\frac{\pi}{14}\right)$$

$\therefore$ Putting $z=\dfrac{1+y}{1-y}$ in (3), we get $\dfrac{\dfrac{(1+y)^7}{(1-y)^7}-1}{\dfrac{1+y}{1-y}+1}$,

$$2^6\cos^2\frac{\pi}{14}\cos^2\frac{3\pi}{14}\cos^2\frac{5\pi}{14}\left(y^2+\tan^2\frac{\pi}{14}\right)$$

$$=\frac{\left(y^2+\tan^2\frac{3\pi}{14}\right)\left(y^2+\tan^2\frac{5\pi}{14}\right)}{(1-y)^6}$$

or, $$\frac{(1+y)^7+(1-y)^7}{2}=2^5\times\frac{7}{64}\left(y^2+\tan^2\frac{\pi}{14}\right)\left(y^2-\tan^2\frac{3\pi}{14}\right)\left(y^2+\tan^2\frac{5\pi}{14}\right)$$

$\therefore$ Equating coefficient of y^4 from both sides, we get

$$\frac{1}{2}\left({}^7C_4\right)=7\left(\tan^2\frac{\pi}{14}+\tan^2\frac{3\pi}{14}+\tan^2\frac{5\pi}{14}\right)$$

$$\therefore\ \tan^2\frac{\pi}{14}+\tan^2\frac{3\pi}{14}+\tan^2\frac{5\pi}{14}=5$$

Example 39:

Consider a 'n' sided regular polygon with origin as it's centre. If z_1 be the complex number representing a vertex A_1, of the polygon, find the complex number associated with the vertex that is adjacent to A_1.

Solution:

Vertex adjacent to A_1 is either A_2 or A_p. Let z_2 and z_1 represent the vertices A_2 and A_n respectively. Considering the rotation about the origin, we get

$$\frac{z_2 - 0}{z_1 - 0} = \frac{|z_2|}{|z_1|} e^{i2\pi/n} \quad = e^{i2\pi/n}$$

$\Rightarrow z_2 = z_1\ e^{2p/n}$

Similarly,

$$\frac{z_n}{z_1} = e^{-2\pi/n}$$

$\Rightarrow z_n = z_1 e^{-i2\pi/n}$.

Example 40:

Consider a square ABCD such that z_1, z_2, z_3 and z_4 represent its vertices A, B, C and D respectively. Express 'z_3' and 'z_4' in terms of z_1 and z_2.

Solution:

Consider the rotation of AB about A through an angle π/4, we get

$$\frac{z_3 - z_1}{z_2 - z_1} = \frac{|z_3 - z_1|}{|z_2 - z_1|} e^{i\pi/4}$$

$$= \sqrt{2}\left(\cos\frac{\pi}{4} + i\sin\frac{\pi}{4}\right)$$

$\Rightarrow \quad z_3 = z_1 + (z_2 - z_1)\,(1 + i)$

Similarly, $\dfrac{z_4 - z_1}{z_2 - z_1} = \dfrac{|z_4 - z_1|}{|z_2 - z_1|} e^{i\pi/2} = i$

$\Rightarrow \quad z_4 = z_1 + i\,(z_2 - z_1)$

Example 41:

Complex numbers z_1, z_2, z_3 are the vertices A,B,C respectively of an isosceles right angled triangle with right angle at C. Show that $(z_1 - z_2)^2 = 2(z_1 - z_3)(z_3 - z_2)$

Solution:

In the isosceles triangle ABC, AC = BC and BC ⊥ AC. It means that AC is rotated through angle π/2 to occupy the position BC.

Hence we have $\dfrac{z_2 - z_3}{z_1 - z_3} = e^{-i\pi/2} = + i$

$\Rightarrow z_2 - z_3 = + i(z_1 - z_3)$

$\Rightarrow z_2^2 + z_3^2 - 2z_2z_3 = -\left(z_2^2 + z_3^2 - 2z_1z_3\right)$

$\Rightarrow z_1^2 + z_2^2\ 2z_1z_2 = 2z_1z_3 + 2z_2z_3 - 2z_1z_2 - 2z_3^2$

$= 2(z_1 - z_3)(z_3 - z_2)$

$\Rightarrow (z_1 - z_2)^2 = 2(z_1 - z_3)(z_3 - z_2).$

Example 42:

Let z_1 and z_2 be the roots of the equation $z^2 + pz + q = 0$, where the coefficients p and q may be complex number. Let A and B represent z_1 and z_2 in the complex plane. If $\angle AOB = a \neq 0$ and $OA = OB$, where O is the origin, prove that $p^2 = 4\theta \cos^2\left(\frac{\alpha}{2}\right)$.

Solution:

$z_1 + z_2 = -p$ ...(1)

and $z_1z_2 = q$ ...(2)

$z_2 = z_1e^{i\alpha}$ ...(3)

Now $p^2 = (z_1 + z_2)^2$

$= (z_1 + z_1e^{i\alpha})^2$

$= z_1^2(1 + 2e^{i\alpha} + e^{i2\alpha})$

$= qe^{-i\alpha}(1 + 2e^{i\alpha} + e^{i2\alpha})$

$= q(2 + e^{-i\alpha} + e^{i\alpha})$

$= q(2 + 2\cos\alpha) = 4q\cos^2\dfrac{\alpha}{2}$

Example 43:

Consider the complex numbers $A(z_1)$, $B(z_2)$ and $C(z_3)$ as the vertices of a triangle ABC. Let P be the fcot of the altitude from A on BC.

Find the complex number associated with the point P.

Solution:

$\because \angle APC = p/2 \Rightarrow \arg\left(\dfrac{z - z_1}{z_2 - z_3}\right) = \dfrac{\pi}{2}$ or $\dfrac{\pi}{2}$

$$\Rightarrow \frac{z - z_1}{z_2 - z_3} \text{ is purely imaginary}$$

$$\Rightarrow \frac{z - z_1}{z_2 - z_3} + \frac{\bar{z} - \bar{z}_1}{\bar{z}_2 - \bar{z}_3} = 0$$

$$\Rightarrow \frac{\bar{z}}{\bar{z}_2 - \bar{z}_3} = \frac{\bar{z}_1}{\bar{z}_2 - \bar{z}_3} - \frac{z - z_1}{z_2 - z_3}$$

Also P(z) is lying on the through B and C.

$$\Rightarrow \frac{z - z_2}{z_2 - z_3} = \frac{\bar{z} - \bar{z}_2}{\bar{z}_2 - \bar{z}_3}$$

$$\Rightarrow \frac{\bar{z}}{\bar{z}_2 - \bar{z}_3} = \frac{\bar{z}_2}{\bar{z}_2 - \bar{z}_3} + \frac{z - z_2}{z_2 - z_3}$$

$$\frac{z - z_2}{z_2 - z_3} + \frac{\bar{z}_2}{\bar{z}_2 - \bar{z}_3} = \frac{\bar{z}_1}{\bar{z}_2 - \bar{z}_3} - \frac{z - z_1}{z_2 - z_3}$$

$$\Rightarrow z = \frac{1}{2}\left(\frac{(\bar{z}_1 - \bar{z}_2)(z_2 - z_3)}{\bar{z}_2 - \bar{z}_3} + z_1 + z_2\right).$$

Example 44:

If $|z| = 1$, *then prove that the points represented by* $\sqrt{\frac{1+z}{1-z}}$ *lie on one or other of two fixed perpendicular straight lines.*

Solution:

Since $|z| = 1$. z lies on a unit circle having centre at the origin. arg $\left(\frac{1+z}{1-z}\right) = +\frac{\pi}{2}$ or $+\frac{3\pi}{2}$

$$\Rightarrow \frac{1+z}{1-z} = ke^{i\pi/2} \text{ or } ke^{i3\pi/2}$$

where k is a real parameter and its value depends upon the position of z.

Let $\alpha = \sqrt{\frac{1+z}{1-z}}$

$\Rightarrow$ $\alpha = \sqrt{k}\ e^{-i\pi/4}$ or $\sqrt{k}\ e^{i3\pi/4}$

$\Rightarrow$ α lies on one or other of the two perpendicular lines.

Example 45:

If $\alpha = \frac{z-i}{z+i}$*, show that, when z lies above the real axis, α will lie within the unit circle which has centre at the origin. Find the locus of α as z travels on the real axis from* $-\infty$ *to* $+\infty$*.*

Solution:

From figure, it is clear that $|z-i| < |z-i|$ (as z lies above the real axis).

$$\Rightarrow |a| = \frac{|z-i|}{|z+i|} < 1$$

$\Rightarrow$ α lies within the unit circle which has centre at the origin.

Now if z is travelling one the real axis Im(z) = 0 and Re(z) varies from $-\infty$ to $+\infty$.

Let $z = x + i0$

$\Rightarrow \alpha = \frac{x-i}{x+i}$ moves on the unit circle which has centre at the origin.

Example 46:

Two different non-parallel lines cut the circle $|z| = r$ *in points a, b, c and d respectively. Prove that these lines meet in the point z given by* $z = \frac{a^{-1}+b^{-1}-c^{-1}-d^{-1}}{a^{-1}b^{-1}-c^{-1}d^{-1}}$.

Solution:

Since P, Q, R are collinear,

$$\begin{vmatrix} z & \bar{z} & 1 \\ c & \bar{c} & 1 \\ d & \bar{d} & 1 \end{vmatrix} = 0$$

$$\Rightarrow z\,z(\bar{c}-\bar{d}) - \bar{z}\ (c-d) + (c\bar{d}-\bar{c}d) = 0 \qquad \text{...(1)}$$

Similarly, $z(\bar{a}-\bar{b}) - \bar{z}(a-b) + (a\bar{b}-\bar{a}b) = 0$ ...(2)

From $\{(1) \times (a-b)\} - \{(2) \times (c-d)\}$

$$z[(\bar{c}-\bar{d})\ (a-b) - (\bar{a}-\bar{b})(c-d)] = (a\bar{b}-b\bar{a})(c-d) - (c\bar{d}-\bar{c}d)(a-b)$$

Now $a\bar{a} = r^2 \Rightarrow \bar{a} = \frac{r^2}{a}$,

Similarly $\bar{b} = \frac{r^2}{b}$, $\bar{c} = \frac{r^2}{c}$, $\bar{d} = \frac{r^2}{d}$ ...(3)

From (3) $z\left[\left(\frac{r^2}{c} - \frac{r^2}{d}\right)(a-b) - \left(\frac{r^2}{a} - \frac{r^2}{b}\right)(c-d)\right]$

$$= \left(\frac{ar^2}{b} - \frac{br^2}{a}\right)(c-d) - \left(\frac{cr^2}{d} - \frac{r^2}{c}d\right)(a-b)$$

$$\Rightarrow z\left[\frac{(d-c)(a-b)}{cd} - \frac{(b-a)(c-d)}{ab}\right]$$

$$= \frac{(a^2-b^2)(c-d)}{ab} - \frac{(c^2-d^2)(a-b)}{cd}$$

$$\Rightarrow z\left[-\frac{1}{cd} + \frac{1}{ab}\right] = \frac{(a+b)}{ab} - \frac{c+d}{cd}$$

$$\Rightarrow z = \frac{a^{-1} + b^{-1} - c^{-1} - d^{-1}}{a^{-1}b^{-1} - c^{-1}d^{-1}}.$$

Example 47:

Plot the region represented by $\frac{\pi}{3} \le \arg\left(\frac{z+1}{z-1}\right) \le \frac{2\pi}{3}$ *in the Argand plane.*

Solution:

Let us take $\arg\left(\frac{z+1}{z-1}\right)$ = 2p/3. Clearly z lies on the minor arc of Circle passing through (1, 0) and (–1, 0). Similarly, $\arg\left(\frac{z+1}{z-1}\right)$ = p/3 means that 'z' is lying on the major arc of the circle passing through (1, 0) and (–1, 0). Now if we take any point in the region included between the two arcs, say $P_1(z_1)$, we get $\frac{\pi}{3} \le \arg\left(\frac{z+1}{z-1}\right) \le \frac{2\pi}{3}$.

Thus, $\frac{\pi}{3} \le \arg\left(\frac{z+1}{z-1}\right) \le \frac{2\pi}{3}$ represents the shaded region (excluding the points (1, 0) and (–1, 0).

Example 48:

If z_1, z_2, z_3 *are complex numbers such that* $\frac{2}{z_1} = \frac{1}{z_2} + \frac{1}{z_3}$, *show that the points represented by* z_1, z_2, z_3 *lie on a circle passing through the origin.*

Solution:

Since $P(z_1)$, $Q(z_2)$, $R(z_3)$ and $S(z_4)$ are concyclic points,

$\Delta PSQ = \angle PRQ$

$$\Rightarrow \arg.\frac{z_2 - z_4}{z_1 - z_4} = \arg\frac{z_2 - z_3}{z_1 - z_3}$$

$$\Rightarrow \arg\left[\left(\frac{z_2 - z_4}{z_1 - z_4}\right)\left(\frac{z_1 - z_3}{z_2 - z_3}\right)\right] = 0$$

$$\Rightarrow \frac{(z_2 - z_4)}{(z_1 - z_4)}.\frac{(z_1 - z_3)}{(z_2 - z_3)} = \text{real} \qquad ...(1)$$

We have $\frac{2}{z_1} = \frac{1}{z_2} + \frac{1}{z_3}$ from which

$$z_3 = \frac{z_1 z_2}{2z_2 - z_1} \qquad ...(2)$$

From (1) and (2), $\frac{z_2}{z_1} \times \frac{z_1 - \frac{z_1 z_2}{2z_2 - z_1}}{z_2 - \frac{z_1 z_2}{2z_2 - z_1}} = \text{real}$

$$\Rightarrow \frac{z_2 - z_1}{2(z_2 - z_1)} = \text{real}$$

$\Rightarrow \frac{1}{2}$ = real, which is true.

Therefore z_1, z_2, z_3 and the origin are concylic.

Alternative Solution

$$\frac{2}{z_1} = \frac{1}{z_2} + \frac{1}{z_3}$$

$$\Rightarrow \frac{1}{z_1} - \frac{1}{z_2} = \frac{1}{z_3} - \frac{1}{z_1}$$

$$\Rightarrow \frac{z_2 - z_1}{z_1 z_2} = \frac{z_1 - z_3}{z_1 z_3}$$

$$\Rightarrow \quad \frac{z_2 - z_1}{z_3 - z_1} = -\frac{z_2}{z_3}$$

$$\Rightarrow \quad \arg\left(\frac{z_2 - z_1}{z_3 - z_1}\right) = \arg\left(-\frac{z_2}{z_3}\right) = \pi + \arg\left(\frac{z_2}{z_3}\right)$$

$$\Rightarrow \quad \alpha = \pi - \beta \Rightarrow \alpha + \beta = \pi$$

$\Rightarrow$ points A, B, C, D are concyclic.

Example 49:

Consider the complex numbers $z_1 = 10 + 6i$ and $z_2 = 4 + 2i$. If $\arg\left(\frac{z - z_1}{z - z_2}\right) = \frac{\pi}{4}$, find the centre and radius of the locus of complex number z.

Solution:

Let $O(z_0)$ be the centre of the circle.

We have $\angle AOB = \frac{\pi}{2}$ .AB $= |z_1 - z_2| = |6 + 4i| = \sqrt{52}$

Let OQ = OB = r $\Rightarrow$ AB = $r\sqrt{2}$

$$\Rightarrow \quad r = \sqrt{26}$$

Also $\frac{z_2 - z_0}{z_1 - z_0} = e^{-ip/2} = -i$

$$\Rightarrow \quad z_2 - z_0 = -i(z_1 - z_0)$$

$$\Rightarrow \quad z_0 = \frac{1}{2}(z_2 - iz_2 + iz_1 + z_1) = 5 - 7i.$$

Alternative Solution:

We have $\frac{\pi}{4} = \arg.\frac{z - z}{z - z_2}$

$$= \arg(z - z_1) - \arg(z - z_2)$$

$$\Rightarrow \tan^{-1}1 = \tan^{-1}\frac{y-6}{x-10} - \tan^{-1}\frac{y-2}{x-4}$$

$$\Rightarrow \quad \tan^{-1} = \tan^{-1}\left[\frac{\frac{y-6}{x-10} - \frac{y-2}{x-4}}{1 + \frac{(y-6)(y-2)}{(x-10)(x-4)}}\right]$$

$$\Rightarrow \qquad 1 = \frac{6y - 4x + 4}{x^2 + y^2 - 14x - 8y + 52}$$

$\Rightarrow \qquad x^2 + y^2 - 10x - 14y + 48 = 0,$

or $\qquad (x - 5)^2 + (y - 7)^2 = 26 = \left(\sqrt{26}\right)^2$

or $\qquad |x - 5 + i(y - 7)| = \sqrt{26}$

$\Rightarrow \qquad |z - 5 - 7i| \ \sqrt{26}\,.$

Example 50:

If 'a' is a complex number such that |a| = 1. Find the values of a, so that equation $az^2 + z + 1 = 0$ has one purely imaginary root.

Solution:

$az^2 + z + 1 = 0$ ...(1)

Taking conjugate of both sides, $az^2 + z + 1 = \overline{0}$

$\Rightarrow \bar{a}(\bar{z})^2 + \bar{z} + \bar{1} = 0$

$\bar{a}(\bar{z})^2 + \bar{z} + \bar{1} = 0$

$\bar{a}\ z^2 - z + 1 = 0$ (Since $\bar{z} = -z$ as z is purely imaginary) ...(2)

Eliminating z from both the equations.

$[(\bar{a} - a)^2 + 2\ (a + \bar{a}\,] = 0$

Let a = cos θ + isin θ (since |a| = 1)

So $(-2i \sin\theta)^2 + 2\ (2 \cos\theta) = 0$

$\Rightarrow -4 \sin^2\theta + 4 \cos\theta = 0$

$\Rightarrow \cos^2\theta + \cos\theta - 1 = 0$

$$\cos\theta = \frac{-1 \pm \sqrt{1+4}}{2}$$

Only feasible value of cos θ is $\dfrac{\sqrt{5} - 1}{2}$

Hence a = cos θ + isin θ, where $\theta = \cos^{-1}\left(\dfrac{\sqrt{5} - 1}{2}\right)$.

Example 51:

Construct an equation whose roots are +i, sec(2π/5), sec(4π/5).

Solution:

Let $5\theta = 2np$, $n \in |$

$\Rightarrow \cos 3\theta = \cos 2\theta$

$\Rightarrow 4\cos^3\theta - 2\cos^2\theta - 3\cos\theta + 1 = 0$

Put $\cos\theta = x$, here note that θ may be 0 or $\frac{2\pi}{5}$ or $\frac{4\pi}{5}$ for n = 0, 1, 2 and also if we take n = 3, 4, 5, ..., then the value of $\cos\theta$ will start repeating.

$\Rightarrow 4x^3 - 2x^2 - 3x + 1 = 0$

$\Rightarrow (x - 1)(4x^2 + 2x - 1) = 0$

$\Rightarrow 4x^2 + 2x - 1 = 0$

where $x = \cos\frac{2\pi}{5}, \cos\frac{4\pi}{5}$.

Hence required equation is

$$\left(\frac{4}{x^2}+\frac{2}{x}-1\right)(x^2 + 1) = 0$$

$\Rightarrow (x^2 - 2x - 4)(x^2 + 1) = 0$

$\Rightarrow x^4 - 2x^3 - 3x^2 - 2x - 4 = 0.$

Example 52:

If $iz^3 + z^3 - z + i = 0$, then show $|z| = 1$.

Solution:

$iz^3 + z^2 - z + i = 0$

By substituting $z = i$ in the equation, we get 0=0

Hence $z - i$ is a factor of $iz^3 + z^3 - z + i$

$\Rightarrow iz^2 (z - i)(z - i) = 0$

$\Rightarrow (iz^2 - 1)(z - i) = 0$

Either $iz^2 - 1 = 0$ or $z - i = 0$

When $z - i = 0$, $z = i$

$\therefore \quad |z| = |0 + i.1| = \sqrt{0^2 + 1^2} = 1\cdot$

When $iz^2 - 1 = 0$, $z^2 = \frac{1}{i} = -i$

$\therefore |z^2| = |0 - 1.i| = \sqrt{0^2 + (-1)^2} = 1$

$\therefore |z^2| = 1$ or $|z| = 1$

∴ In any case we have $|z| = 1$.

Example 53:

ABCD is a rhombus. The diagonals AC and BD intersect at the point M and satisfy BD = 2AC. Its points D and M represent the complex numbers 1 + i and 2 – i respectively. Find the complex number represented by A.

Solution:

Let A be (x, y)

It is given that BD = 2AC ⇒ MD = 2AM

Also DM is perpendicular to AM

$$\Rightarrow \quad (1-2)^2 + (1+1)^2$$

$$= \quad 4\,[(x-2)^2 + (y+1)^2] \qquad \text{...(1)}$$

and $\quad \dfrac{y+1}{x-2} \cdot \dfrac{1+1}{1-2} = -1$

$$\Rightarrow \quad 2(y+1) = x - 2$$

With $x - 2 = 2(y + 1)$, (1) gives $(y+1)^2 = 1/4$

$$\Rightarrow \quad y = -1/2, -3/2 \Rightarrow x = 3, 1$$

⇒ A represents $z = 3 - i/2$, or $1 - 3i/2$

Alternative Solution:

MD = 2AM and AM ⊥ DM *i.e.,* AMD = π/2

$$\Rightarrow \quad \frac{z-(2-i)}{(1+i)-(2-i)} = \frac{AM}{MD}.e^{\pm\frac{i\pi}{2}} = \pm\, i/2$$

$$\Rightarrow \quad z - (2-i) = \pm\frac{i}{2}(-1+2i)$$

$$\Rightarrow \quad z = 3 - i/2 \text{ or } 1 - 3i/2.$$

Example 54:

For every value of c ≥ 0, find all the complex numbers z which satisfy the equation $|z|^2 - 2iz + 2c(1+i) = 0$

Solution:

If $z = x + iy$, then the given equation reduces to

$$x^2 + y^2 - 2ix + 2y + 2c + 2ic = 0 \Rightarrow x = c,$$

so that the real part yields $y^2 + 2y + c^2 + 2c = 0$.

$\Rightarrow \quad y \neq R^+$

For real y, $1 - 2c - c^2 \geq 0$

$\Rightarrow \quad c^2 + 2c - 1 \leq 0.$

$\Rightarrow \quad (c + \sqrt{2} + 1)\,[c - (\sqrt{2} - 1)] \leq 0 \leq c \leq \sqrt{2} - 1.$

Hence the solutions $c + \left[-1 \pm \sqrt{1-2c-c^2}\right] i \; 0 \leq c \leq \sqrt{2}-1$.

For c = 0, y = 0, – 2, so that z = 0, z = – 2i.

Example 55:

Find the complex numbers z which simultaneously satisfy the equations

$$\left|\frac{z-12}{z-8i}\right| = \frac{5}{3}, \left|\frac{z-4}{z-8}\right| = 1.$$

Solution:

Now $\left|\frac{z-4}{z-8}\right| = 1 \Rightarrow \left|\frac{x-4+iy}{x-8+iy}\right| = 1$

$\Rightarrow \quad (x - 4)^3 + y^2 = (x - 8)^2 + y^2$

$\Rightarrow \quad x = 6.$

With x = 6, $\left|\frac{z-12}{z-8i}\right| = \frac{5}{3} \Rightarrow \left|\frac{-6+iy}{6+i(y-8)}\right| = \frac{5}{3}$

$\Rightarrow \quad 9(36 + y^2) = 25[36 + (y - 8)^2]$

$\Rightarrow \quad y^2 - 25y + 136 = 0$

$\Rightarrow \quad y = 178.$

Hence, the required numbers are

$$z = 6 + 17i,\ 6 + 8i.$$

Example 56:

Prove the following inequalities geometrically and analytically:

(a) $\left|\frac{z}{|z|} - 1\right| \leq |\arg z|$

(b) $|z-1| \leq |z| - 1 + |z|\,|\arg z|$.

Solution:

(a) The quantity z/|z| lies on a unit circle centred at the origin.

It is clear that $\left|\frac{z}{|z|}-1\right|$ = AP ≤ arc(AP)=a=arg z.

If a is negative, this gives $\left|\frac{z}{|z|}-1\right| \le |\arg z|$.

To show this analytically, put z = r (cos α + i sin α), where r = |z|. Then

$$\left|\frac{z}{|z|}-1\right|$$

$$= |(\cos\alpha - 1) + i \sin\alpha|$$
$$= [(\cos\alpha - 1)^2 + \sin^2\alpha]^{1/2}$$
$$= [2(1 - \cos\alpha)]^{1/2}$$
$$= \left(4\sin^2\frac{\alpha}{2}\right)^{1/2} = 2\left|\sin\frac{\alpha}{2}\right|$$
$$\le 2\left|\frac{\alpha}{2}\right| \quad \therefore |a| = |\arg z|.[\sin\theta \le \theta \text{ if } \theta \ge 0]$$

(b) Referring to the shown figure and using the result of part (a), we have (z – 1) = AP ≤ QP = ||z| – 1| + |z – |z||

$$\le ||z| - 1| + |z| \left|\frac{z}{|z|}-1\right| \le ||z| - 1| + |z|\,|\arg z|.$$

Analytically, we can get this result as follows:

$$|z - 1| = |(z - |z|) + (|z| - 1) \le |z-|z|| + ||z| + 1|$$

$$\Rightarrow |z - 1| \le ||z| - 1| + |z| \left|\frac{z}{|z|}-1\right| \le ||z| - 1| + |z|\,|\arg z|$$

Example 57:

Find the equation of the circle which touches the line iz + $\bar{z}$ *+ 1 + i = 0 and has the lines (1 – i) z = (1 + i)* $\bar{z}$ *and (1 + i) z + (1 – i)* $\bar{z}$ *= 4i as its normals.*

Solution:

Clearly point of intersection of normals would be the centre of the required circle.

$$(1 - i)\, z = (1 + i)\, \bar{z}$$

$$\Rightarrow \quad \bar{z} = \frac{1-i}{1+i} z$$

$$(1 + i)\, z + (i - 1) \bar{z} = 4i$$

$$\Rightarrow \quad \bar{z} = \frac{4i-(1+i)z}{(1-1)}$$

$$\frac{1-i}{1+i} z = \frac{4i-(1+i)z}{(i-1)}$$

$$\Rightarrow \quad (1 - i)\,(i - 1)\, z = 4i\,(1 + i) - (1 + i)^2 z$$

$$\Rightarrow \quad z(1 + i)^2 - (1 - i)^2 = 4i\,(1 + i)$$

$$\Rightarrow \quad z(4i) = 4i\,(1 + i)$$

$$z = (1 + i)$$

Now equation of tangent can be rewritten as,

$$\frac{1}{(1-i)}(iz+\bar{z})+1 = 0$$

$$\Rightarrow \quad i(1 - i)\, z + (1 - i)\, \bar{z} + 2 = 0$$

$$\Rightarrow \quad (i + 1)z + (1 - i) \bar{z} + 2 = 0$$

Now distance of $z = (1 + i)$ from this line

$$= \frac{|(1+i)\,(1+i) + (1-i)\,(1-i) + 2|}{2|1-i|} = \frac{1}{\sqrt{2}}$$

Thus the equation of required circle is,

$$|z - (1 + i) = \frac{1}{\sqrt{2}}.$$

Example 58:

Find $a \in R$ if atleast one complex number z is to satisfy $|z + 3) = a^2 - 2a + 6$ and $|z - 3\sqrt{3}i| < a^2$ simultaneously.

Solution:

$$|z + 3| = a^2 - 2a + 1 \qquad ...(1)$$

$\Rightarrow$ z lies on a circle whose centre is $(- 3 + i.0)$

and radius is $a^2 - 2a + 6$. Here $a^2 - 2a + 6 > 0$, $\perp a \in R$ (as the roots of the corresponding equation are imaginary)

And $|z - 3\sqrt{3}i| < a^2$ ...(2)

$\Rightarrow$ z lies in the interior of a circle whose centre is $(0 + 3\sqrt{3}i)$ and radius is a^2.

Clearly, atleast one complex would satisfy both the equations if the two circles cut in two real and distinct points or first circle lies entirely inside the second circle.

We know that two circles with centres at c_1 and c_2 and radii being r_1 and r_2 respectively, will never intersect if $c_1c_2 \geq r_1 + r_2$

$\Rightarrow |-3 - 3\sqrt{3}i| \geq a^2 + a^2 - 2a + 6$

$\Rightarrow 6 \geq 2a^2 - 2a + 6 \Rightarrow a^2 - a \leq 0$

$\Rightarrow a \in [0, 1]$ ...(3)

Hence desired values of a is $R - (-\infty\ 1]$

$= (1, \infty)$.

Example 59:

Consider a triangle formed by the points $A\left(\frac{2}{\sqrt{3}}e^{i\frac{\pi}{2}}\right), B\left(\frac{2}{\sqrt{3}}e^{i\frac{\pi}{6}}\right)$, $C\left(\frac{2}{\sqrt{3}}e^{i\frac{5\pi}{6}}\right)$. *Let P(z) be any point on it's in circle. Prove that* $AP^2 + BP^2 + CP^2 = 5$.

Solution:

Let $z_1 = \frac{2}{\sqrt{3}}e^{i\frac{\pi}{2}}, z_2 = \frac{2}{\sqrt{3}}e^{-i\frac{\pi}{6}}, z_3 = \frac{2}{\sqrt{3}}e^{-i\frac{5}{6}}$.

Clearly the points lie on the circle $|z| = \frac{2}{\sqrt{3}}$

If 'l' be the length of side of the ΔABC.

$AD = |\sin 60^\circ =| \frac{\sqrt{3}}{2}$

$\Rightarrow \quad OD = \frac{1}{3}AD = \frac{1\sqrt{3}}{6}$

Now, $\quad OA = \frac{2}{3}AD = \frac{2}{3}.\frac{1\sqrt{3}}{2} = \frac{2}{\sqrt{3}}$

$\Rightarrow \qquad |= 2$ and $OD = \dfrac{2\sqrt{3}}{6} = \dfrac{1}{\sqrt{3}}$

$\Rightarrow$ Equation of in circle is $|z| = \dfrac{1}{\sqrt{3}}$

Let P(z) be any point on the in circle,

$\Rightarrow \qquad z = \dfrac{1}{\sqrt{3}} e^{i\pi}$

Now, $\quad AP^2 = |z - z_1|^2 = |z|^2 + |z_1|^2 - \left(z\bar{z}_1 + \bar{z}z_1\right)$

Similarly, $BP^2 = |z|^2 + |z_2|^2 - \left(z\bar{z}_2 + \bar{z}z_2\right)$

and $\quad CP^2 = |z|^2 + |z_3|^2 - \left(z\bar{z}_3 + \bar{z}z_3\right)$

$$AP^2 + BP^2 + CP^2 = 3|z|^2 + |z_1|^2 + |z_2|^2 + |z_3|^2$$

$$z\left(\bar{z}_1 + \bar{z}_2 + \bar{z}_3\right) - \bar{z}\,(z_1 + z_2 + z_3)$$

$$= 1 + 3.\frac{4}{3} - z(0) - \bar{z}(0) = 5.$$

Example 60:

Locate the complex number z = x + iy for which

(i) $\log_{1/2} |z - 2| \log_{1/2} |z|$

(ii) $\log \log_{1/\sqrt{3}} \dfrac{|z|^2 - z + 1}{2 + |z|} > -2$

(iii) $\log_{14}\left(13 + |z^2 - 4i|\right) + \log_{196} \dfrac{1}{\left(13 + |z^2 + 4i\right)^2} = 0$

(iv) $\log_{1/2} \dfrac{|z-1|+4}{3|z-1|-2} > 1$

(v) $\log_{\cos\frac{\pi}{6}} \dfrac{|z-2|+5}{4|z-2|-4} < 2$

(vi) $\log_{\sec\frac{\pi}{4}} \dfrac{|z|^2 + |z| + 4}{2|z| - 1} > 2$.

Solution:

$$\log_{\frac{\sqrt{3}}{2}} \frac{|z-2|+5}{4|z-2|-4} < \log_{\frac{\sqrt{3}}{2}} \left(\frac{\sqrt{3}}{2}\right)^2$$

so $\dfrac{|z-2|+5}{4|z-2|-4} > \dfrac{3}{4}$ $\left(\text{since base } \dfrac{\sqrt{3}}{2} < 1\right)$

Also for the logarithmic to be defined

$4|z-2| - 4 > 0$, *i.e.*, $|z-2| > 1$

when $|z-2| > 1$, $\dfrac{|z-2|+5}{4|z-2|-4} > \dfrac{3}{4}$

gives $4(|z-2| + 5) > 3.4(|z-2| - 1)$

or $|z-2| + 5 > 3|z-2| - 3$

or $8 > 2|z-2|$

$\Rightarrow |z-2| < 4$

$\Rightarrow$ z is such that $1 < |z-2| < 4$

Hence, the region for z in the Argand plane is as shown being by the shaded area. This the area shaded between two circles with a common centre z = 2 and radii 1 and 4.

Example 61:

(i) *Prove that*

$$\left(a_1^2 + b_1^2\right)\left(a_2^2 + b_2^2\right) \ldots \left(a_n^2 + b_n^2\right)$$

can written as the sum of the two squares.

(ii) *Let $x_1, x_2, x_3, \ldots, x_n$ be the roots of the equation $x^n + x^{n-1} + \ldots + x + 1 = 0$ compute the expression*

$$\frac{1}{x_1-1} + \frac{1}{x_2-1} + \ldots + \frac{1}{x_n-1}.$$

(iii) *If $|z_1| = |z_2| = \ldots = |z_n| = 1$,*

prove that $|z_1 + z_2 + z_3 + \ldots + z_n|$

$$= \left|\frac{1}{z_1} + \frac{1}{z_2} + \frac{1}{z_3} + \ldots + \frac{1}{z_n}\right|.$$

Solution:

(iii) $|z_r| = 1$ for r = 1 to n

$\Rightarrow z_r\bar{z}_r = 1 \Rightarrow \bar{z}_r = \dfrac{1}{z_r}$

Let $a = z_1 + z_2 + \ldots z_n$

$$\Rightarrow \bar{\alpha} = \bar{z}_1 + \bar{z}_2 + \ldots + \bar{z}_n = \left(\frac{1}{z_1} + \frac{1}{z_2} + \ldots + \frac{1}{z_n}\right) \text{ as } |\alpha| = |\bar{\alpha}|$$

$$\Rightarrow |z_1 + z_2 + \ldots + z_n| = \left|\frac{1}{z_1} + \frac{1}{z_2} + \ldots + \frac{1}{z_n}\right|$$

Example 62:

(i) *Let z be a non-real complex number lying on the circle* $|z| = 1$*. then prove that*

$$z = \frac{1 + i\tan\left(\frac{\arg z}{2}\right)}{1 - i\tan\left(\frac{\arg z}{2}\right)}.$$

(ii) *Find the modulus and the argument of the complex number* z_1*, where* $z_1 = z^2 - z$ *and* $z = \cos\theta + i\sin\theta$*.*

(iii) *If* z_1 *and* z_2 *are complex numbers, prove that*

$|z_1 + z_2|^2 + |z_1|^2 + |z_2|^2$ *if and only if* $z_1\bar{z}_2$ *is purely imaginary.*

Solution:

(i) Since z lies on $|z| = 1$

$z = \cos\alpha + i\sin\alpha,\ \alpha = \arg(z)$

$$= \frac{\cos^2\frac{\alpha}{2} - \sin^2\frac{\alpha}{2} + 2i\sin\frac{\alpha}{2}\cos\frac{\alpha}{2}}{\cos^2\frac{\alpha}{2} + \sin^2\frac{\alpha}{2}}$$

$$= \frac{\left(\cos\frac{\alpha}{2} + i\sin\frac{\alpha}{2}\right)^2}{\left(\cos\frac{\alpha}{2} - i\sin\frac{\alpha}{2}\right)\left(\cos\frac{\alpha}{2} + i\sin\frac{\alpha}{2}\right)}$$

$$= \frac{\cos\frac{\alpha}{2} + i\sin\frac{\alpha}{2}}{\cos\frac{\alpha}{2} - i\sin\frac{\alpha}{2}} = \frac{1 + i\tan\frac{\alpha}{2}}{1 - i\tan\frac{\alpha}{2}} = \frac{1 + i\tan\left(\frac{\arg z}{2}\right)}{1 - i\tan\left(\frac{\arg z}{2}\right)}$$

(ii) $z_1 = z^2 - z = (z - 1)$

$\Rightarrow |z_1| = |z|\,|z - 1|$ and $\arg(z_1) = \arg(z) + \arg(z - 1)$

$z = e^{i\theta} \Rightarrow z - 1 = e^{i\theta} - 1 = 2i \sin \frac{\theta}{2} e^{i\theta/2}$

$\Rightarrow |z_1| = 2\ |\sin \theta/2|$

and $\arg(z_1) = \theta + \arg\left(2i \sin \frac{\theta}{2} e^{i\theta/2}\right) = \theta + \left(\frac{\pi}{2} + \frac{\theta}{2}\right) = \frac{1}{2}(\pi + 3\theta)$

(if $z \neq 1$)

If $z = 1$, $|z_1| = 1$ and $\arg(z_1|$ will not be defined.

Example 63:

(i) *If the complex slope of the line joining the points represented by* z_1, z_2 *in the Argand plane is defined as* $\frac{z_1 - z_2}{\bar{z}_1 - \bar{z}_2}$, *find the condition for perpendicular or two lines with complex slopes,* ω_1 *and* ω_2.

(ii) *If* $x_1 \leq x_2$ *and* $y_1 \leq y_2$ *for complex numbers* $z_1 = x_1 + iy_1$ *and* $z_2 = x_2 + iy_2$ *we write* $z_1 \leq z_2$. *Show that all complex numbers z with* $1 \leq z$ *we have* $\frac{1-z}{1+z} \leq 0$.

Solution:

(i) Let $\omega_1 = \frac{z_1 - z_2}{\bar{z}_1 - \bar{z}_2}$ and $\omega_2 = \frac{z_3 - z_4}{\bar{z}_3 - \bar{z}_4}$

Now $|\omega_2| = \left|\frac{z_1 - z_2}{\bar{z}_1 - \bar{z}_2}\right| = \left|\frac{z_1 - z_2}{z_1 - z_2}\right| = 1 \quad \left(\text{as} \left|\frac{z}{\bar{z}}\right| = 1\right)$

Similarly $|\omega_2| = 1$

Now amp ω_1 = amp $\frac{z_1 - z_2}{\bar{z}_1 - \bar{z}_2}$ = amp $\frac{z_1 - z_2}{z_1 - z_2}$

= 2 amp $(z_1 - z_2)$ ...(1)

Similarly, amp ω_2 = 2 amp $(z_3 - z_4)$ ...(2)

Now amp$\frac{z_1 - z_2}{z_3 - z_4} = \pm\frac{\pi}{2}$ because the line segments joining z_1, z_2 and z_3, z_4 are at right angle

From (1) an (2), we get

amp ω_1 = amp ω_2 = 2 {amp$(z_1 - z_2)$ – amp $(z_3 - z_4)$

= 2 amp$\frac{z_1 - z_2}{z_3 - z_4} = 2\left(\pm\frac{\pi}{2}\right) = \pm\pi$

$$\Rightarrow \text{amp}\frac{\omega_1}{\omega_2} = \pm\pi \Rightarrow \frac{\omega_1}{\omega_2} = \left|\frac{\omega_1}{\omega_2}\right|$$

$\{\cos(\pm p) + i\sin(\pm\pi)\} = -1 = \omega_1 + \omega_2 = 0.$

(ii) Let $z = x + iy$ the $1 \le z \Rightarrow 1 \le x$ and $0 \le y$

Consider $z' = \dfrac{1-z}{1+z}$

$$\text{Re}(z') = \text{Re}\left(\frac{1-z}{1+z}\right) = \text{Re}\left(\frac{1-x-iy}{1+x+iy}\right)$$

$$= \text{Re}\left(\frac{(1-x-iy)(1+x-iy)}{(1+x)^2 + y^2}\right)$$

$$= \frac{(1-x)(1+x)-y^2}{(1+x)^2+y^2} = \frac{1-x^2-y^2}{(1+x)^2+y^2} \le 0$$

$$\text{Im}(z') = \text{Im}\left(\frac{1-z}{1+z}\right) = \frac{(1-x)(x-y)+(-y)(1+x)}{(1+x)^2+y^2}$$

$$= \frac{-2y}{(1+x)^2+y^2} \le 0$$

Thus, we have $\text{Re}(z') \le 0$ and $\text{Im}(z') \le 0$

$\Rightarrow z' \le 0.$

Example 64:

z_1, z_2, z_3 are three non-zero complex numbers such that z_2 ¹ z_1, and $a = |z_1|$, $b = |z_2|$, $c = |z_3|$. If $\begin{vmatrix} a & b & c \\ b & c & a \\ c & a & b \end{vmatrix} = 0$, *then show that*

$$\textit{arg.}\frac{z_3}{z_2} = \arg\left(\frac{z_3 - z_1}{z_2 - z_1}\right)^2.$$

Solution:

$$\begin{vmatrix} a & b & c \\ b & c & a \\ c & a & b \end{vmatrix} = (a+b+c)(a-b)(b-c)(c-a) = 0$$

$\Rightarrow a = b = c$ (as $a + b + c \neq 0$)

$\Rightarrow z_1, z_2$ and z_3 lie on a circle with centre at origin

Clearly $\angle BOC = 2\ \angle BAC$

$$\Rightarrow \arg\left(\frac{z_3}{z_2}\right) = 2\arg\left(\frac{z_3 - z_1}{z_2 - z_1}\right) = \arg\left(\frac{z_3 - z_1}{z_2 - z_1}\right)^2.$$

Example 65:

Find the complex number z with maximum and minimum possible values of $|z|$ satisfying

(i) $z - \frac{1}{z} = 1$;

(ii) $\left|z + \frac{4}{z}\right| = 3$.

Solution:

(ii) $\left|z + \frac{4}{z}\right| = 3 \Rightarrow \left||z| - \frac{4}{|z|}\right| \leq 3$

$\Rightarrow -3 \leq |z| - \frac{4}{|z|} \leq \Rightarrow 1 \leq |z| \leq 4$

$\Rightarrow |z|_{max} = 4$ and $|z|_{min} = 1$

Let us find out the complex number with maximum value of $|z|$

$|z| = 4 \Rightarrow z = 4e^{i\alpha}$

$\Rightarrow |4e^{-i\alpha} + e^{i\alpha}| = 3$

$\Rightarrow |4e^{i\alpha} + e^{i\alpha}|^2 = 9$

$\Rightarrow e^{i2\alpha} + e^{i2\alpha} = -2$

$\Rightarrow \cos 2\alpha = -1 \Rightarrow \alpha = \pm \pi/2$

$\Rightarrow z = 4e^{\pm i\pi/2} = \pm 4i$

Similarly, complex number with minimum value of $|z|$ can be obtained.

Example 66:

(i) *Three points represented by complex numbers a, b, c lie on a circle with centre origin and radius r. The tangents at c cuts the chord joining the points a, b at z. Show that* $z = \frac{a^{-1} + b^{-1} - 2c^{-1}}{a^{-1}b^{-1} - c^{-2}}$.

(ii) *Two points represented by complex numbers a, b lie on a circle with centre Q and radius r. The tangents at 'a' and 'b' intersects at z. Find the complex number representing z.*

Solution:

Equation of line AB

$$\frac{z-z}{a-b}=\frac{\bar{z}-\bar{z}}{\bar{a}-\bar{b}}=\frac{\bar{z}-\frac{r^2}{a}}{r^2\left(\frac{1}{a}-\frac{1}{b}\right)}$$

$$\Rightarrow \frac{z-a}{-b}=\frac{\left(a\bar{z}-r^2\right)}{r^2}\Rightarrow \frac{\bar{z}}{r^2}=\left(1-\frac{z-a}{b}\right)\frac{1}{a} \qquad ...(1)$$

Equation of tangent at C(c) is :

$$z\bar{c}+\bar{z}c=2r^2$$

$$\Rightarrow c\frac{\bar{z}}{r^2}=2-\frac{z}{c}$$

$$\Rightarrow \frac{\bar{z}}{r^2}=\frac{1}{c}\left(2-\frac{z}{c}\right) \qquad ...(ii)$$

Eliminate $\frac{\bar{z}}{r_2}$ from (i) and (ii)

Example 67:

A is represented by the number 6i, B by the number 3, and P by z. If PA = 2PB, then show that the locus of P is the circle $z\bar{z}=(4+2i)z+(4-2i)\bar{z}$. *Find its radius and the centre.*

Solution:

$A \equiv 6i,\quad B \equiv 3, P \equiv P\ (z)$

PA = 2PB

$\Rightarrow |z-6i| = 2\,|z-3|$

$\Rightarrow (z-6i)\left(\bar{z}+6i\right) = 4(z-3)\left(\bar{z}-3\right)$

$\Rightarrow z\bar{z} = (4+2i)\,z + (4-2i)\,\bar{z}$

Example 68:

(i) *Show that the triangle whose vertices are* z_1, z_2, z_3 *and a.b.c. are*

similarly, if $\begin{vmatrix} z_1 & a & 1 \\ z_2 & b & 1 \\ z_3 & c & 1 \end{vmatrix}=0$.

(ii) *If a, b, c and u, v, w are the complex numbers representing the vertices of two triangles such that c = (1 – r) a + rb, w = (1 – r) u + rv, where r is non real complex then prove that the triangles are similar.*

Solution:

(i) The triangle are similar, therefore

$$\frac{z_1 - z_2}{z_3 - z_2} = \frac{a-b}{c-b}$$

$$\Leftrightarrow \begin{vmatrix} z_1 - z_2 & a-b \\ z_3 - z_2 & c-b \end{vmatrix} = 0$$

$$\Leftrightarrow \begin{vmatrix} z_1 - z_2 & a-b & 0 \\ z_2 & b & 1 \\ z_3 - z_2 & c-b & 0 \end{vmatrix} = 0 \Leftrightarrow \begin{vmatrix} z_1 & a & 1 \\ z_2 & b & 1 \\ z_3 & c & 1 \end{vmatrix} = 0$$

(ii) $c = (1 - r)\, a + rb,\ \omega = (1 - r)\, u + rv$

$\Rightarrow c\text{-}a = r\,(b - a),\ \omega\text{-}u = r(v - u)$

$$\Rightarrow r = \frac{c-a}{b-a},\ r = \frac{\omega - u}{v-u}$$

$$\Rightarrow \frac{c-a}{b-a} = \frac{\omega - u}{v-u} \Rightarrow \frac{|c-a|}{|b-a|} = \frac{|\omega - u|}{|v-u|}$$

and $\arg\left(\dfrac{c-a}{b-a}\right) = \arg\left(\dfrac{\omega - u}{v-u}\right)$

Hence, given triangles would be similar.

Example 69:

Prove that the area of the triangle whose vertices are the points represented by the complex numbers z_1, z_2, z_3 on the Arand diagram is $\sum\left(\dfrac{(z_2 - z_3)|z_1|^2}{4iz_1}\right)$.

Solution:

$$\text{Area of } \Delta ABC = \frac{1}{2}.AB.CD$$

$$= \frac{1}{2}|z_1 - z_2|CD$$

Equation of AB is $z\bar{a}+\bar{z}a+b$ = where

$$a = i(z_1 - z_1),\ b = \left(\frac{\bar{z}_1 z_2 - z_1\bar{z}_2}{i}\right)$$

$$\Rightarrow \text{CD} = \frac{|z_3\bar{a} + \bar{z}_3 a + b|}{2|a|}$$

$$\Rightarrow \text{Required area} = \frac{1}{4|a|}|z_1 - z_2|\,|z_3\bar{a}+\bar{z}_3 a+b|\cdot$$

Example 70:

(i) *If the equation $ax^2 + bx + c = 0$, $(0 < a < b < c)$ has non-real complex roots z_1 and z_2, show that $|z_1| > 1$, $|z_2| > 1$.*

(ii) *If z is a complex number and $z^2 + az + b=0$ $(a = 0)$ has roots both of which have unit modulus then prove that $|a| \le 2$, $|b| =1$, arg (b) = 2arg(a).*

Solution:

(i) $ax^2 + bx + c = 0$ $(c > b > a > 0)$

As z_1 and z_2 are non real roots of the equation

$\Rightarrow$ $z_1 = \bar{z}_2$

we have $z_1 z_2 = \dfrac{c}{a} > 1$

$\Rightarrow$ $|z_2|^2 > 1 \Rightarrow |z_2| > 1$

$\Rightarrow$ $|z_2| > 1$. Similarly $|z_1| > 1$.

Example 71:

(i) *If z_1, z_2 are two complex numbers representing consecutive vertices a regular hexagon then find the complex number z_3 representing the vertex adjacent to z_2.*

(ii) *The three points z_1, z_2, z_3 are connected by the relation $az_1 + bz_1 + cz_3 = 0$, where a.b.c are real and $a + b + c = 0$. Prove that the three points are collinear.*

Solution:

(i) Clearly $|z_1 - z_2| = |z_3 - z_2|$ ...(1)

and amp $\dfrac{z_1 - z_2}{z_3 - z_2} = \pm\dfrac{2\pi}{3} = \angle Z_1Z_2Z_3$

(where the point Z_1 represents z_1 etc.)

$$\Rightarrow \frac{z_1 - z_2}{z_3 - z_2} = \frac{|z_1 - z_2|}{|z_3 - z_2|}\left\{\cos\frac{2\pi}{3} + \pm i \sin\frac{2\pi}{3}\right\}$$

$$= \left\{-\frac{1}{2} \pm i\frac{\sqrt{3}}{2}\right\}, \text{ using (1)}$$

$\therefore z_1 - z_2 = a(z_3 - z_2)$,

(where $a = \dfrac{-1 \pm i\sqrt{3}}{2} = \omega, \omega^2$ (and $a^3 = 1$)

or $a^2 (z_1 - z_2) = a^3 (z_3 - z_2) = z_3 - z_2$

or $z_3 = z_2 + a^2 (z_1 - z_2) = a^2 z_1 + (1 - a^2)z_2$

$$= \left(-\frac{1}{2} \pm \frac{\sqrt{3}}{2}\right)z_2 + \left(1 - \frac{-1 \pm i\sqrt{3}}{2}\right)z_2$$

$$= -\frac{1}{2}\left(1 \pm i\sqrt{3}\right)z_1 + \frac{1}{2}\left(3 \pm \sqrt{3}\right)z_2$$

Note: If z_1, z_2, z_3 are given to be in anticlock wise sense then amp $\dfrac{z_1 - z_2}{z_3 - z_2} = \dfrac{2\pi}{3}$ will be taken.

Example 72:

(i) *Prove that* $|2z \cos a + z^2| < 1$, *if* $|z|$ £ 0.41.

(ii) *If* $|z - 3i| < \sqrt{5}$ *then prove that the complex number z also satisfies the inequality* $|i(z + 1) + 1| < 2\sqrt{5}$.

(iii) *Find the complex number z which satisfies the condition* $|z - a + ai| = 1$ *and having the greatest absolute value, where a is a real constant.*

Solution:

(ii) Here $i(z + 1) + 1$ has to be written as the sum of two complex numbers containing $z - 3i$ because we have to use $|z - 3i| < \sqrt{5}$

$|i(z + 1) + 1| = |i(z - 3i) + i - 2|$ £ $|i(z - 3i)| + |-2 + i)$

(by triangle inequality)

$= |i|\ |z - 3i| + \sqrt{5} < \sqrt{5} + \sqrt{5} = 2\sqrt{5}$

(iii) Let z – a + ai = cos θ + i sin θ

$\Rightarrow \quad z = a + \cos\theta + i(\sin\theta - a)$

$$\Rightarrow |z| = \sqrt{2a^2 + 1 + 2\sqrt{2}a\cos\left(\theta + \frac{\pi}{4}\right)}$$

Case I: A > 0, |z| is maximum if θ = – π/4

$\therefore \quad |z| = \sqrt{2}\,a + 1$

and $\quad z = \left(\sqrt{2}a + 1\right)\left(\frac{1}{\sqrt{2}} - \frac{i}{\sqrt{2}}\right)$

Case II: A < 0, |a| is maximum if θ = π – π/4 = 3π/4

$\therefore \quad |z| = \left|\sqrt{2}a - 1\right|$

$\therefore \quad z = \left|\sqrt{2}a - 1\right|\left(-\frac{1}{\sqrt{2}} + \frac{i}{\sqrt{2}}\right).$

2

Exponential, Trigonometric and Hyperbolic Functions of a Complex Variable

(Separation into Real and Imaginary Parts)

2.1 THE EXPONENTIAL FUNCTIONS OF A COMPLEX VARIABLE

We know that if x is any real number, then

$$e^x = 1 + \frac{x}{1!} + \frac{x^2}{2!} + \frac{x^3}{3!} + \text{... ad.infi.}$$

We shall take motivation from this expansion of e^x as a power series in x, to define the exponential function e^z of a complex variable $z = x + iy$, where x and y are real. Thus we define

$$e^z = 1 + \frac{z}{1!} + \frac{z^2}{2!} + \frac{z^3}{3!} + \text{... ad.infi.} \qquad \text{...(1)}$$

It can be shown by D' Alembert's ratio test, that the power series (1) is absolutely convergent for all values of the complex variable z. If we denote the nth term of the series (1) by u_n, we have

$$\lim_{n\to\infty} \cdot \left|\frac{u_{n+1}}{u_n}\right| = \lim_{n\to\infty} \left|\frac{z^n}{n!} \cdot \frac{(n-1)!}{z^{n-1}}\right|$$

$$= \lim_{n\to\infty} \frac{|z|}{n} = 0,$$

which is < 1. Therefore the series (1) is absolutely convergent and hence convergent for all values of the complex variable z. Thus the sum of the series (1) exists for all z and this justifies our definition of e^z given in (1). Hence for the complex variable z, we define e^z as the sum of the power series (1).

The other ways to denote e^x are E(z) and exp(z). Either of these symbol is read a 'exponential z'.

Putting z = 0 on both sides of (1), we see that $e^0 = 1$.

Remark: It should be noted that when z is complex, e^z is only y a symbol used to represent the sum of the series on the R.H.S. of (1). It does not mean that

$$e^z = \left(1+\frac{1}{1!}+\frac{1}{2!}+\frac{1}{3!}+\ldots+\infty\right)^z.$$

2.2 INDEX LAW FOR THE EXPONENTIAL FUNCTIONS

To prove that $e^{z}{}_{1}.e^{z}{}_{2} = e^{z}{}_{1} + z_2$

i.e., $E(z_1).E(z_2) = E(z_1 + z_2)$.

Proof: If z_1 and z_2 be two complex number, then by the definition of E(z), we have

$$E(z_1) = 1+\frac{z_2}{1!}+\frac{z_2^2}{2!}+\frac{z_2^3}{3!}+\ldots+\frac{z_2^n}{n!}+\ldots\infty \qquad \ldots(1)$$

and $$E(z_2) = 1+\frac{z_1}{1!}+\frac{z_1^2}{2!}+\frac{z_1^3}{3!}+\ldots+\frac{z_1^n}{n!}+\ldots\infty \qquad \ldots(2)$$

Since the series (1) and (2) are absolutely convergent, therefore by Cauchy's theorem on multiplication of absolutely convergent series, we have

$$E(z_1) \,.\, E(z_2) = \left(1+\frac{z_1}{1!}+\frac{z_1^2}{2!}+\ldots+\frac{z_1^n}{n!}+\ldots\right)\left(1+\frac{z_2}{1!}+\frac{z_2^2}{2!}+\ldots+\frac{z_2^n}{n!}+\ldots\right)$$

$$= 1+\frac{1}{1!}(z_1+z_2)+\left(\frac{z_1^2}{2!}+\frac{z_1 z_2}{1!}+\frac{z_2^2}{2!}\right)+\ldots$$

$$+\left(\frac{z_1^n}{n!}+\frac{z_1^{n-1}}{(n-1)!}\cdot\frac{z_2}{1!}+\frac{z_1^{n-2}}{(n-2)!}\cdot\frac{z_2^2}{2!}+\ldots\frac{z_2^n}{n!}\right)+\ldots$$

$$= 1+\frac{1}{1!}(z_1+z_2)+\frac{1}{2!}\left(z_1^2+2z_1z_2+z_2^2\right)+\ldots$$

$$+\frac{1}{n!}\left[z_1^n+n\,z_1^{n-1}z_2+\frac{n(n-1)}{2!}z_1^{n-2}z_2^2+\ldots+z_2^n\right]+\ldots$$

$$= 1+\frac{(z_1+z_2)}{1!}+\frac{(z_1+z_2)^2}{2!}+\ldots+\frac{(z_1+z_2)^n}{n!}+\ldots$$

$= E(z_1 + z_2)$, by def. of $E(z)$.

The above result may also be written as

$$\exp(z_1).\exp(z_2) = \exp(z_1 + z_2).$$

Deductions:

(i) If $z_1, z_2, \ldots, z_n$ be n complex number, then from the above law, we see by induction that

$E(z_1).E(z_2) \ldots E(z_n) = E(z_1 + z_2 + \ldots + z_n)$.

If we put $z_1 = z_2 = \ldots = z_n = z$ (say), we have

$[E(z)]^n = e^{nz}$ or $[\exp(z)]^n = \exp(nz)$.

(ii) If z be any complex number, then

$e^z . e^{-z} = e^{z-z} = e^0 = 1$.

From this result, it follows that $e^z \neq 0$ for all z.

For if there were such a value of z, then since e^{-z} would exist for this value of z, we would have $0 = 1$, which is absurd.

2.3 TRIGONOMETRICAL FUNCTIONS OR CIRCULAR FUNCTIONS OF A COMPLEX VARIABLE

We know that if x is any real number, then

$$\cos x = 1 - \frac{x^2}{2!} + \frac{x^4}{4!} - \ldots + (-1)^n \frac{x^{2n}}{(2n)!} + \cdots \text{ad.inf.}$$

and $$\sin x = 1 - \frac{x^3}{3!} + \frac{x^5}{5!} - \ldots + (-1)^n \frac{x^{2n+1}}{(2n+1)!} + \cdots \text{ad.inf.}$$

We shall take motivation from these expansions of cos x and sin x as power series in x, to define cos z and sin z, when z is a complex number.

So if $z = x + iy$ is a complex variable, we define

$$\cos z = 1 - \frac{z^2}{2!} + \frac{z^4}{4!} - \ldots + (-1)^n \frac{z^{2n}}{(2n)!} + \ldots \text{ad.inf.}$$

$$\cos z = 1 - \frac{z^2}{2!} + \frac{z^4}{4!} - \ldots + (-1)^n \frac{z^{2n}}{(2n)!} + \ldots \text{ad.inf.}$$

and $$\sin z = z - \frac{z^3}{3!} + \frac{z^5}{5!} - \ldots + (-1)^n \frac{z^{2n+1}}{(2n+1)!} + \ldots \text{ad.inf.}$$

Our definitions are sensible because both the series used to define cos z and sin z are absolutely convergent for all z.

Replacing z by –z in the above definitions, we find that

cos (–z) = cos z, and sin (–z) = – sin z.

The other circular functions of a complex variable are defined in the same way as those for a real variable. Thus we define

$$\tan z = \frac{\sin z}{\cos z}, \quad \cot z = \frac{\cos z}{\sin z}, \quad \sec z = \frac{1}{\cos z} \text{ and } \operatorname{cosec} z = \frac{1}{\sin z}.$$

Remark: For a complex variable z = x + iy, we have defined e^z, cos z, sin z etc. in such a way that if we take

z = x + i0 + x (*i.e.*, real),

then these definitions give results which are in sound agreement with those for a real variable.

2.4 EULER'S THEOREM

If θ be real or complex, we have $e^{i\theta} = \cos\theta + i\sin\theta$.

Proof:

If z is any complex number, then by definition of e^z, we have

$$e^z = 1 + \frac{z}{1!} + \frac{z^2}{2!} + \frac{z^2}{3!} + \ldots$$

Replacing z by iθ, we have

$$e^{i\theta} = 1 + \frac{i\theta}{1!} + \frac{(i\theta)^2}{2!} + \frac{(i\theta)}{3!} + \frac{(i\theta)^4}{4!} + \ldots$$

$$= \left(1 - \frac{\theta^2}{2!} + \frac{\theta^4}{4!} - \ldots\right) + i\left(\theta - \frac{\theta^3}{3!} + \frac{\theta^5}{5!} - \ldots\right)$$

= cos θ + i sin θ, by definitions of cos θ and sin θ.

Thus $e^{i\theta} = \cos\theta + i\sin\theta$. ...(1)

This result is known a *Euler's theorem.*

Replacing θ by –θ in (1), we get

$$e^{-i\theta} = \cos(-\theta) + i\sin(-\theta)$$

or $e^{-i\theta} = \cos\theta - i\sin\theta$. ...(2)

From (1) and (2) on addition and subtraction, we get

$$e^{i\theta} + e^{-i\theta} = 2\cos\theta,$$

and $e^{i\theta} - e^{-i\theta} = 2i\sin\theta$.

$$\therefore \quad \cos\theta = \frac{e^{i\theta}+e^{-i\theta}}{2} \text{ and } \sin\theta = \frac{e^{i\theta}-e^{-i\theta}}{2i} \qquad ...(3)$$

These results are known as *Eulers exponential values of* cos θ and sin θ.

From these exponential values of cos θ and sin θ, we see that

$$\tan\theta = \frac{\sin\theta}{\cos\theta} = \frac{e^{i\theta}-e^{-i\theta}}{i\left(e^{i\theta}+e^{-i\theta}\right)}, \ \cot\theta = i\frac{e^{i\theta}+e^{-i\theta}}{e^{i\theta}-e^{-i\theta}}, \text{ etc.}$$

Putting θ = 0 in the formulae (3), we get

$$\cos 0 = \frac{1}{2}\left(e^0+e^0\right) = \frac{1}{2}(1+1) = \frac{1}{2}\cdot 2 = 1, \text{ and } \sin 0 = 0.$$

2.5 PERIODICITY OF FUNCTIONS

(a) Prove that e^z is a periodic function, where z is a complex quantity.

Proof: If z = x + iy, then

$$e^z = e^{x+iy} = e^x.e^{iy}$$

$$= e^x(\cos y + i\sin y), \text{ by Euler's theorem}$$

$$= e^x[\cos(2n\pi + y) + i\sin(2n\pi + y)], \text{ where n is any integer}$$

$$= e^x.e^{i(2n\pi + y)} = e^{x+iy+2n\pi i} = e^{z+2n\pi i}.$$

Hence e^z is a periodic function of period 2πi.

(b) Prove that sin z, tan z etc. are periodic functions, where z is a complex quantity.

Proof: We have

$$\cos(z + 2n\pi) = \cos z\cos 2n\pi - \sin z\sin 2n\pi$$

$$= \cos z, \text{ n being any integer}$$

$$\sin(z + 2n\pi) = \sin z\cos 2n\pi + \cos z\sin 2n\pi$$

$$= \sin z, \text{ n being any integer}$$

and

$$\tan(z + n\pi) = \frac{\sin(z+n\pi)}{\cos(z+n\pi)}$$

$$= \frac{\sin z\cos n\pi + \cos z\sin n\pi}{\cos z\cos n\pi - \sin z\sin n\pi}$$

$$= \tan z, \text{ n being any integer.}$$

From these we conclude that cos z and sin z are periodic functions of period 2π and tan z is a periodic function of period π.

2.6 DE MOIVRE'S THEOREM FOR COMPLEX ARGUMENT

We have already mentioned De Moivre's theorem for real values of θ. Now we extend that theorem for complex values of θ. We have

$(e^{i\theta})^n = e^{in\theta}$, whether θ be real or complex.

$\therefore$ $(\cos\theta + i\sin\theta)^n = \cos n\theta + i\sin n\theta$, by Euler's theorem.

This shows that De-Moivre's Theorem is true whether θ be real complex.

2.7 SOME STANDARD TRIGONOMETRICAL RESULTS FOR COMPLEX ARGUMENTS

Since our trigonometric functions of a complex variable are generalizations of trigonometric functions of a real variable, therefore most of our results for trigonometric functions of a real variable also hold good for trigonometric functions of a complex variable.

To prove that for all values of x, y, real or complex, the following are true:

(i) $\cos^2 x + \sin^2 x = 1$,

(ii) $\cos(-x) = \cos x$,

(iii) $\cos 2x = \cos^2 x - \sin^2 x$,

(iv) $\sin 3x = 3\sin x - 4\sin^3 x$,

(v) $\cos 3x = 4\cos^3 x - 3\cos x$,

(vi) $\sin x + \sin y = 2\sin\frac{1}{2}(x+y)\cos\frac{1}{2}(x-y)$,

(vii) $\cos x - \cos y = 2in\frac{1}{2}(x+y)\sin\frac{1}{2}(y-x)$

(viii) $\sin(x \pm y) = \sin x\cos y \pm \cos x\sin y$,

(ix) $\cos(x \pm y) = \cos x\cos y \pm \sin x\sin y$.

Proof: (i) To prove that $\cos^2 x + \sin^2 x = 1$

$$\text{L.H.S.} = \cos^2 x + \sin^2 x$$

$$= \left\{\frac{e^{ix}+e^{-ix}}{2}\right\} + \left\{\frac{e^{ix}-e^{-ix}}{2i}\right\}^2$$

$$= \frac{1}{4}\left[\left(e^{ix}+e^{-ix}\right)^2 - \left(e^{ix}-e^{-ix}\right)^2\right] \qquad [\because i^2 = -1]$$

$$= \frac{1}{4}[(e^{2ix} + e^{-2ix} + 2e^{ix}.e^{-ix}) - (e^{2ix} + e^{-2ix} - 2e^{ix}.e^{-ix})$$

$= \frac{1}{4}[4e^{ix}.e^{-ix}] = e^{ix-ix} = e^0 = 1 =$ R.H.S.

Aliter: We have $\cos^2 x + \sin^2 x = (\cos x + i \sin x)(\cos x - i \sin x)$

(ii) To prove that cos (–x) = cox x

$$\text{L.H.S.} = \cos(-x) = \frac{e^{i(-x)} + e^{-i(-x)}}{2}$$

$$= \frac{e^{-ix} + e^{ix}}{2} = \frac{e^{ix} + e^{-ix}}{2} = \cos x = \text{R.H.S.}$$

(iii) To prove that cos 2x = cos²x – sin²x

R.H.S. $= \cos^2 x - \sin^2 x$

$$= \left\{\frac{e^{ix} + e^{-ix}}{2}\right\}^2 - \left\{\frac{e^{ix} - e^{-ix}}{2i}\right\}^2,$$

by Euler's exponential values of cos x and sin x

$$= \frac{1}{4}\left[\left(e^{ix} + e^{-ix}\right)^2 + \left(e^{ix} - e^{-ix}\right)^2\right] \qquad [\because i^2 = -1]$$

$$= \frac{1}{4}\left[\left(e^{2ix} + e^{-2ix} + 2e^{ix}.e^{-ix}\right) + \left(e^{2ix} + e^{-2ix} - 2e^{ix}.e^{-ix}\right)\right]$$

$$= \frac{1}{4}\left[2\left(3e^{2ix} + e^{-2ix}\right)\right] = \frac{1}{2}\left(e^{2ix} + e^{-2ix}\right)$$

$$= \frac{1}{2}\left[e^{i(2x)} + e^{-i(2x)}\right] = \cos 2x = \text{L.H.S.}$$

(iv) To prove that

sin 3x 3 sin x – 4 sin³x

$$\text{L.H.S.} = \sin 3x = \frac{e^{i(3x)} - e^{-(3x)}}{2i} = \frac{e^{3ix} - e^{-3ix}}{2i} \qquad \textbf{(Note)}$$

$[\because a^3 - b^3 = (a - b)^3 + 3ab(a - b)]$

$$= \frac{1}{2i}\left[(2i \sin x)^3 + 3.e^0.(2i \sin x)\right] \qquad \left[\because \sin x = \frac{e^{ix} - e^{-ix}}{2i}\right]$$

$$= \frac{1}{2i}\left[8i^3 \sin^3 x + 6i \sin x\right] = \frac{1}{2i}\left[2i\left(-4\sin^3 x + 3\sin x\right)\right]$$

(v) To prove that cos

3x = 4 cos³ x = 3 cos x

$$\text{L.H.S.} = \cos 3x = \frac{1}{2}\left[e^{-i(3x)}\right] = \frac{1}{2}\left[\left(e^{ix}\right)^3 + \left(e^{-ix}\right)^3\right]$$

$$= \cos 3x = \frac{1}{2}\left[e^{-i(3x)}\right] = \frac{1}{2}\left[\left(e^{ix}\right)^3 + \left(e^{-ix}\right)^3\right]$$

$$= \frac{1}{2}\left[\left(e^{ix} + e^{-ix}\right)^3 - 3e^{ix}.e^{-ix}\left(e^{ix} + e^{-ix}\right)\right]$$

$$[\because a^3 + b^3 = (a + b)^3 - 3ab\ (a + b)]$$

$$= \frac{1}{2}\left[(2\cos x)^3 = 3.e^0.(2\cos x)\right]$$

$$= \frac{1}{2}\left[(2\cos x)^3 = 3.e^0.(2\cos x)\right] \qquad \left[\because \cos x = \frac{1}{2}\left(e^{ix} + e^{-ix}\right)\right]$$

$$= \frac{1}{2}\left[8\cos^3 x - 6\cos x\right] = 4\cos^3 x - 3\cos x = \text{R.H.S.}$$

(iv) To prove that $\sin x + \sin y = 2\sin\frac{1}{2}(x + y)\cos\frac{1}{2}(x - y)$

L.H.S. = sin x + sin y

$$= \frac{e^{ix} - e^{-ix}}{2i} + \frac{e^{iy} - e^{-iy}}{2i} = \frac{1}{2i}\left[\left(e^{ix} - e^{-ix}\right) + \left(e^{iy} - e^{-iy}\right)\right]$$

$$= \left(\frac{1}{2i}\right)\ [\{e^{i[(x + y)/2 + (x - y)]} - e^{-i[(x + y)/2 + (x - y)/2]}\}$$

$$+ \{e^{i[(x + y)/ - (x - y)/2]} - e^{-i[(x + y)/2 - (x - y)/2]}\}]\ \textbf{(Note)}$$

$$= \left(\frac{1}{2i}\right)\ [\{e^{i[(x + y)/2 + (x - y)/2]} + e^{i[(x + y)/2 - (x - y)/2]}$$

$$- \{e^{-i[(x + y)/2 + (x - y)/2]} + e^{-i[(x + y)/2 - (x - y)/2]}\}]\ \textbf{(Note)}$$

$$= \left(\frac{1}{2i}\right)\ [e^{i(x + y)/2}\ \{e^{i(x - y)/2} + e^{-i(x - y)/2}\}$$

$$- e^{-i\,(x + y)/2}\ \{e^{-i(x - y)/2} + e^{i(x - y)/2}\}]$$

$$= \frac{1}{2i}\ [e^{i(x - y)/2} + e^{-i(x - y)/2}]\ [e^{i(x + y)/2} - e^{-i(x + y)/2}]$$

$$= \frac{1}{2i}\left[2\cos\left\{\frac{x - y}{2}\right\}\right]\cdot\left[2i\ \sin\left\{\frac{x + y}{2}\right\}\right]$$

$$= 2\sin\frac{1}{2}(x + y)\cos\frac{1}{2}(x - y) = \text{R.H.S.}$$

(vii) To prove that

$$\cos x - \cos y = 2\sin\frac{1}{2}(x+y)\cdot\sin\frac{1}{2}(y-x)$$

$$\text{R.H.S.} = 2\sin\frac{1}{2}(x+y)\cdot\sin\frac{1}{2}(y-x)$$

$$= 2\left[\frac{e^{i(x+y)/2} - e^{-i(x+y)/2}}{2i}\right]\left[\frac{e^{i(y-x)/2} - e^{-i(y-x)/2}}{2i}\right]$$

$$= -\frac{1}{2}\left[e^{i(x+y+y-x)/2} - e^{i(x+y-y+x)/2} - e^{-i(x+y-y+x)/2} + e^{-i(x+y+y-x)/2}\right]$$

(simply by multiplication)

$$= -\frac{1}{2}\left[e^{iy} - e^{ix} - e^{-ix} + e^{-iy}\right] = \frac{1}{2}\left(e^{ix} + e^{-ix}\right) - \frac{1}{2}\left(e^{iy} + e^{-iy}\right)$$

(viii) To prove that sin (x + y) = sin x :os y + cos x sin y

$$\text{R.H.S.} = \frac{e^{ix} - e^{-ix}}{2i} \times \frac{e^{iy} + e^{-iy}}{2} + \frac{e^{ix} + e^{-ix}}{2} \times \frac{e^{iy} - e^{-iy}}{2i}$$

$$= \left(\frac{1}{4i}\right)\ [\{e^{i(x+y)} + e^{i(x-y)} - e^{-i(x-y)}\}$$

$$+ \{e^{i(x+y)} - e^{i(x-y)} + e^{-i(x-y)} - e^{-i(x+y)}\}]$$

$$= \frac{1}{4i}\left[2e^{i(x+y)} - 2e^{-i(x+y)}\right] = \frac{e^{i(x+y) - e^{-i(x+y)}}}{2i}$$

$= \sin(x + y) =$ L.H.S.

Similarly, $\sin(x - y) = \sin x \cos y - \cos x \sin y$.

$\therefore \sin(x \pm y) = \sin x \cos y \pm \cos x \sin y$.

(ix) To prove that cos (x ± y) = cos x cos y ± sin x sin y

We have $\cos x \cos y + \sin x \sin y$

$$= \frac{e^{ix} + e^{-ix}}{2} \cdot \frac{e^{iy} + e^{-iy}}{2} + \frac{e^{ix} - e^{-ix}}{2i} \cdot \frac{e^{iy} - e^{-iy}}{2i}$$

$$= \frac{1}{4}[e^{i(x+y)} + e^{i(x+y)} + e^{-i(x-y)} + e^{-i(x+y)}]$$

$$-\frac{1}{4}[e^{i(x+y)} - e^{i(x-y)} - e^{-i(x-y)} + e^{-i(x+y)}] \quad [\because i^2 = -1]$$

$$= \frac{1}{4}\left[2e^{i(x-y)} + 2e^{-i(x-y)}\right] = \frac{1}{2}\left[e^{i(x-y)} + e^{-i(x-y)}\right] = \cos(x-y).$$

Similarly we can prove that

$$\cos x \cos y - \sin x \sin y = \cos (x + y).$$

$$\therefore \quad \cos (x \pm y) = \cos x \cos y \pm \sin x \sin y.$$

Remark: If u and v are any complex numbers, we can easily prove that

$$2 \sin u \cos v = \sin (u + v) + \sin (u - v),$$

$$2\cos u \sin v = \sin (u + v) - \sin (u - v)$$

$$2 \cos u \cos v = \cos (u + v) + \cos (u - v),$$

$$2 \sin u \sin v = \cos (u - v) - \cos (u + v)$$

2.8 HYPERBOLIC FUNCTIONS

Definition: For all values of x, real or complex, the quantity $\frac{e^x + e^{-x}}{2}$ is called the *hyperbolic cosine* of x and is written as *cosh x*. Similarly, the quantity $\frac{e^x - e^{-x}}{2}$ is called the *hyperbolic sine* of x and is written as *sinh x* (read as shine x). Thus, we define

$$\mathbf{sinhx = \frac{e^x - e^{-x}}{2} \quad and \quad coshx = \frac{e^x + e^{-x}}{2}.}$$

The hyperbolic tangent, contangent, secant and cosecant are obtained from hyperbolic cosine and sine just as circular tangent, contangent, secant and cosecant are obtained from circular cosine and sine. Thus

$$\tanh x = \frac{\sinh x}{\cosh x} = \frac{\left(e^x - e^{-x}\right)/2}{\left(e^x + e^{-x}\right)/2} = \frac{e^x - e^{-x}}{e^x + e^{-x}}$$

$$\coth x = \frac{\cosh x}{\sinh x} = \frac{\left(e^x + e^{-x}\right)/2}{\left(e^x - e^{-x}\right)/2} = \frac{e^x + e^{-x}}{e^x - e^{-x}}$$

$$\operatorname{sec hx} = \frac{1}{\cosh x} = \frac{1}{\left(e^x + e^{-x}\right)/2} = \frac{2}{e^x + e^{-x}}$$

and $$\operatorname{cosechx} = \frac{1}{\sinh x} = \frac{1}{\left(e^x - e^{-x}\right)/2} = \frac{2}{e^x - e^{-x}}$$

2.9 RELATIONS BETWEEN HYPERBOLIC AND CIRCULAR FUNCTIONS

The hyperbolic functions can be expressed in terms of circular functions as follows:

$$\sin(ix) = \frac{e^{i(ix)} - e^{-i(ix)}}{2i} = \frac{e^{i^2x} - e^{-i^2x}}{2i} = \frac{e^{-x} - e^{x}}{2i}$$

$$= i.\frac{e^{-x} - e^{x}}{2i^2} = i.\frac{e^{x} - e^{-x}}{2}, \qquad [\because i^2 = -1]$$

$$= \mathbf{i\ sinhx.}$$

From this relation sin (ix) = i sinhx,

we have $\sinh x = \frac{1}{i}\sin(ix) = \frac{i}{i^2}\sin(ix) = -i\sin(ix)$.

Similarly $\cos(ix) = \frac{e^{i(ix)} + e^{-i(ix)}}{2} = \frac{e^{i^2x} + e^{-i^2x}}{2} = \frac{e^{-x} + e^{x}}{2} = \frac{e^{x} + e^{-x}}{2}$

$$= \mathbf{coshx.}$$

Now $\tan(ix) = \frac{\sin(ix)}{\cos(ix)} = \frac{i \sinh x}{\cosh x} = i \tanh x$,

cot (ix) $= \frac{1}{\tan(ix)} = \frac{1}{i \tanh x} = \frac{-i}{\tanh x} = -i$ cothx,

sec (ix) $= \frac{1}{\cos(ix)} = \frac{1}{\cosh x} =$ sechx,

and cosec (ix) $= \frac{1}{\sin(ix)} = \frac{1}{i \sinh x} = \frac{-i}{\sinh x} = -$ cosechx.

Similarly we can prove that

sinh (ix) = i sin x; cosh (ix) = cos x;

tanh (ix) = i tan x and coth (ix) = – i cot x, etc.

2.10 PROPERTIES OF HYPERBOLIC FUNCTIONS

Formulae involving hyperbolic functions can be easily deduced with the help of the above relations. A list of some such formulae is given below:

(a) $\sinh 0 = 0$, $\cosh 0 = 1$, $\tanh 0 = 0$.

(b) $\cosh^2 x - \sinh^2 x = 1$

(c) $\sinh 2x = 2\sinh x \cos x$

(d) $\cosh 2x = \cosh^2 x + \sin^2 x = 1 + 2 \sinh^2 x = 2 \cosh^2 x - 1$

(e) $sech^2 = 1 - \tanh^2 x$

(f) $\tanh 2x = 2 \tanh x/(1 + \tanh^2 x)$

(g) (i) $\sinh(x + y)) = \sinh x \cosh y + \cosh x \sinh y$,

(ii) $\cosh(x + y) = \cosh x \cosh y + \sinh x \sinh y$.

(h) $e^x = \cosh x + \sinh x$ and $e^{-x} = \cosh x - \sinh x$.

(i) $\sinh 3x = 3 \sinh x + 4 \sinh^3 x$

(j) $\cosh 3x = 4 \cosh^3 x - 3 \cosh$

(k) $\tanh 3x = \dfrac{3\tanh x + \tanh^3 x}{1 + 3\tanh^2 x}$.

Proof : (a) sinh 0 = 0, cosh 0 = 1, tanh 0 = 0

By definition,

$\sinh x = \frac{1}{2}\left(e^x - e^{-x}\right)$ and $\cosh x = \frac{1}{2}\left(e^x + e^{-x}\right)$.

$\therefore \sinh 0 = \frac{1}{2}\left(e^0 - e^0\right) = \frac{1}{2}(1-1) = 0$ $[\because e^0 = e^{-0} = 1]$

and $\cosh 0 = \frac{1}{2}\left(e^0 + e^{-0}\right) = \frac{1}{2}(1+1) = \frac{1}{2}.2 = 1$

Also $\tanh 0 = \frac{\sinh 0}{\cosh 0} = \frac{0}{1} = 0$.

(b) cosh² x – sinh² x = 1

We have $\cosh^2 x - \sinh^2 x = \left(\frac{e^x + e^{-x}}{2}\right)^2 - \left(\frac{e^x - e^{-x}}{2}\right)^2$

$= \frac{1}{4}\left(e^{2x} + 2 + e^{-2x}\right) - \frac{1}{4}\left(e^{2x} - 2 + e^{-2x}\right) = \frac{1}{2} + \frac{1}{2} = 1$.

(c) sinh 2x = 2 sinhx cosh x

The L.H.S. $= \sinh 2x = \left(\frac{1}{i}\right) \sin(2ix)$ **(Note)**

$= \left(\frac{1}{i}\right) 2\sin(ix)\cos(ix) = \left(\frac{1}{i}\right) 2.(i\sinh x).(\cosh x)$

$= 2 \sinh x \cosh x =$ R.H.S.

(d) cosh 2x = cosh²x + sinh²x = 1 + 2 sinh²x = 2 cosh²x – 1

We have $\cosh 2x = \cos(2ix) = \cos 2(ix)$

$= \cos^2(ix) - \sin^2(ix) = \cosh^2 x - i^2 \sinh^2 x$

$= \cosh^2 x = \sinh^2 x$ ...(1)

$= 1 + \sinh^2 x + \sinh^2 x = 1 + 2\sinh^2 x.$ ...(2)

Also $\cosh 2x = \cosh^2 x + \sinh^2 x = \cosh^2 x + (\cosh^2 x - 1)$

$= 2\cosh^2 x - 1.$ ...(3)

(e) $\text{sech}^2 x = 1 - \tanh^2 x$

We have $\cosh^2 x - \sin^2 x = 1$, by property (b).

Dividing by $\cosh^2 x$, we get $1 - \tanh^2 x = \text{sech}^2 x$.

Similarly, $\coth^2 x - 1 = \text{cosech}^2 x$.

(f) $\tanh 2x = 2\tanh x/(1 + \tanh^2 x)$

We have $i \tanh 2x = \tan(2ix)$.

$\therefore \quad \tanh 2x = \left(\frac{1}{i}\right) \tan(2ix) = \left(\frac{1}{i}\right) \tan 2(ix)$

$$= \frac{1}{i} \cdot \frac{2\tan ix}{\left(1 - \tan^2 ix\right)} = \frac{1}{i} \cdot \frac{2i \tanh x}{\left(1 - i^2 \tanh^2 x\right)} \quad \textbf{(Note)}$$

$$= \frac{2\tanh x}{\left(1 + \tanh^2 x\right)}.$$

Similarly, we find

$$\mathbf{\sinh 2x = \frac{2\tanh x}{1 - \tanh^2 x} \text{ and } \cosh 2x = \frac{1 + \tanh^2 x}{1 - \tanh^2 x}}.$$

(g) (i) $\sinh(x + y) = \sinh x \cosh y + \cosh x \sinh y$

The R.H.S. $= \frac{1}{2}\left(e^x - e^{-x}\right) \cdot \frac{1}{2}\left(e^y + e^{-y}\right) + \frac{1}{2}\left(e^x + e^{-x}\right) \cdot \frac{1}{2}\left(e^y - e^{-y}\right)$

$= \frac{1}{4}[e^{x+y} + e^{x-y} - e^{-(x-y)} - e^{-(x+y)} + e^{x+y} - e^{x-y} + e^{-(x-y)} - e^{-(x+y)}]$

$= \frac{1}{4}\left[2e^{(x+y)} - 2e^{-(x+y)}\right] = \frac{1}{2}\left[e^{(x+y)} - e^{-(x+y)}\right]$

$= \sinh(x + y) =$ L.H.S.

(ii) $\cosh(x + y) = \cosh x \cosh y + \sinh x \sinh y$.

L.H.S. $= \cosh(x + y) = \cos i(x + y) = \cos(ix + iy)$

$= \cos(ix)\cos(iy) - \sin(ix)\sin(iy)$

$= \cosh x \cosh y - i \sinh x \cdot i \sinh y$

$= \cosh x \cosh y - i^2 \sinh x \sinh y$

$= \cosh x \cosh y + \sinh x \sinh y =$ R.H.S.

(h) $e^x = \cosh x + \sinh x$ and $e^{-x} = \cosh x - \sinh x$

We have $\cosh x = \frac{1}{2}\left(e^x + e^{-x}\right)$; $\sinh x = \frac{1}{2}\left(e^x - e^{-x}\right)$.

Adding, we get

$$\cosh x + \sinh x = \frac{1}{2}\left(e^x + e^{-x} + e^x - e^{-x}\right) = \frac{1}{2}.2e^x = e^x$$

and subtracting, we get

$$\cosh x - \sinh x = \frac{1}{2}\left(e^x + e^{-x} - e^x + e^{-x}\right)$$

$$= \frac{1}{2}.2e^{-x} = e^{-x}.$$

(i) $\sinh 3x = 3 \sinh x + 4 \sinh^3 x$

L.H.S. $= \sinh 3x = \left(\frac{1}{i}\right) \sin (3ix) = \left(\frac{1}{i}\right) \sin 3(ix)$

$= \left(\frac{1}{i}\right) [3 \sin (ix) - 4 \sin^3 (ix)]$

$= \left(\frac{1}{i}\right) [3 (i \sinh x) - 4 (i \sinh x)^3]$ [$\because \sin (ix) = i \sinh x$]

$= \left(\frac{1}{i}\right) [3i \sinh x + 4i \sinh^3 x]$ [$\because i^2 = -1; i^2 = -i$]

$= 3 \sinh x + 4 \sinh^3 x =$ R.H.S.

(j) $\cosh 3x = 4 \cosh^3 x - 3 \cosh x$

L.H.S. $= \cosh 3x = \cos (3ix) = \cos 3(ix)$

$= 4 \cos^3 (ix) - 3 \cos (ix)$

$= 4 \cosh^3 x - 3 \cosh x.$ [$\because \cos (ix) = \cosh x$]

(k) $\tanh 3x = (3 \tanh x + \tan^3 x)/(1 + 3 \tanh^2 x)$

L.H.S. $= \tanh 3x = \left(\frac{1}{i}\right) \tan (3ix)$

$$= \left(\frac{1}{i}\right)\left[\frac{3\tan(ix) - \tan^3(ix)}{1 - 3\tan^2(ix)}\right]$$

$$= \frac{1}{i}\left[\frac{3i \tanh x - i^3 \tanh^3 x}{1 - 3i^2 \tanh^2 x}\right]$$ [$\because \tan (ix) = i \tanh x$]

$$= \frac{1}{i}\left[\frac{3i\tanh x + i\tanh^3 x}{1+3\tanh^2 x}\right] = \frac{3\tanh x + \tanh^3 x}{1+3\tanh^2 x}$$

= R.H.S.

2.11 EXPANSIONS IN SERIES FOR sinh x AND cosh x

We have $\sinh x = \frac{1}{2}\left[e^x - e^{-x}\right]$

$$= \frac{1}{2}\left[\left(1+x+\frac{x^2}{2i}+\frac{x^3}{3!}+...\right)-\left(1-x+\frac{x^2}{2!}-\frac{x^3}{3!}+...\right)\right]$$

$$= \frac{1}{2}\left[2x+\frac{2x^3}{3!}+\frac{2x^5}{5!}+...\right] = x+\frac{x^3}{3!}+\frac{x^5}{5!}+...\text{ad.inf.}$$

Thus $\mathbf{\sinh x = x + \frac{x^3}{3!}+\frac{x^5}{5!}+...\ ad.inf.}$

Also $\cosh x = \frac{1}{2}\left[e^x + e^{-x}\right]$

$$= \frac{1}{2}\left[\left(1+x+\frac{x^2}{2!}+\frac{x^3}{3!}+...\right)+\left(1-x+\frac{x^2}{2!}-\frac{x^3}{3!}+...\right)\right]$$

$$= \frac{1}{2}\left[2x+\frac{2x^2}{2!}+\frac{2x^4}{4!}+...\right] = 1+\frac{x^2}{2!}+\frac{x^4}{4!}+...\text{ad.inf.}$$

Thus $\mathbf{\cosh x = 1 + \frac{x^2}{2!}+\frac{x^4}{4!}+...ad.inf.}$

Remark: The students should note that to change a formula for the circular functions into a formula for the hyperbolic functions, we must replace cos x by cosh x and sin x by i sinh x *i.e.,*, we must write $\cosh^2 x$ for $\cosh^2 x$ and $-\sinh^2 x$ for $\sin^2 x$ etc.

2.12 PERIODS OF HYPERBOLIC FUNCTIONS

Be Euler's theorem, we know that

$e^{2n\pi i} = \cos 2n\pi + i \sin 2n\pi = 1 + i.0 = 1$, n being any integer

and $e^{-2n\pi i} = \cos 2n\pi - i \sin 2n\pi = 1 - i.0 = 1$.

$\therefore$ $\sinh(x + 2n\pi i) = \frac{1}{2}[e^{x + 2n\pi i} - e^{-(x + 2n\pi i)}]$, by definition

$= \frac{1}{2}[e^{x} e^{2n\pi i} - e^{-x} e^{-2n\pi i}]$

$= \frac{1}{2}[e^{x} - e^{-x}]$ $\quad$ [$\because e^{\pm 2n\pi i} = 1$, n being any integer]

$= \sinh x.$

Also, $\cosh(x + 2n\pi i) = \frac{1}{2}[e^{x + 2n\pi i} + e^{-(x + 2n\pi i)}]$ $\quad$ [By definition

$= \frac{1}{2}[e^{x}.e^{2n\pi i} + e^{-x}.e^{-2n\pi i}]$

$= \frac{1}{2}[e^{x} + e^{-x}]$ $\quad$ [$\because e^{\pm 2n\pi i} = 1$]

$= \cosh x.$

It follows that sinh x and coshx are periodic functions and both have a period $2\pi i$.

Hence, hyperbolic functions differ from circular functions in having no real period. Their periods are imaginary.

Again, $\tanh(x + n\pi i) = \frac{\sinh(x + n\pi i)}{\cosh(x + n\pi i)}$

$$= \frac{\frac{1}{2}\left[e^{(x+n\pi i)} - e^{-(x+n\pi i)}\right]}{\frac{1}{2}\left[e^{(x+n\pi i)} + e^{-(x+n\pi i)}\right]}$$

$$= \frac{e^{n\pi i}\left[e^{x} - e^{-x}.e^{-2n\pi i}\right]}{e^{n\pi i}\left[e^{x} + e^{-x}.e^{-2n\pi i}\right]}$$

$$= \frac{e^{n\pi i}\left[e^{x} - e^{-x}.e^{-2n\pi i}\right]}{e^{n\pi i}\left[e^{x} + e^{-x}.e^{-2n\pi i}\right]} \quad \textbf{(Note)}$$

$$= \frac{\left[e^{x} - e^{-x}\right]}{\left[e^{x} + e^{-x}\right]} \quad [\because e^{-2n\pi i} = 1]$$

$$= \frac{\sinh x}{\cosh x} = \tanh x.$$

Thus, tanh $(x + n\pi i) = \tanh x$,

i.e., tanh x is a periodic function with period πi.

Therefore, sinh x and cosh x are periodic functions of period $2\pi i$ and tanh x is a periodic function of period πi.

Remark: Note that the hyperbolic functions have imaginary periods while the circular functions have real periods.

2.13 SEPARATION INTO REAL AND IMAGINARY PARTS

By separation or resolving into real and imaginary parts, we mean to put a complex quantity in the form x + iy where x and y are real quantities.

When we are given a fraction to separate into real and imaginary parts whose numerator and denominator are both complex functions, we multiply the numerator and denominator by the conjugate of the denominator and thus make the denominator real.

The conjugate complex is obtained by putting –i for i, *i.e.,* x + iy x + iy and x – iy are conjugate complex quantities.

Now we shall separate the functions e^z, sin z, cos z, tan z, etc. into real and imaginary parts.

(i) To *separate e^z into real and imaginary parts.*

We have $e^z = e^{x+iy} = e^x e^{iy}$

$= e^x (\cos y + i \sin y)$ [By Euler's theorem]

$= e^x \cos y + ie^x \sin y$,

which is of the form a + ib, where a and b are real.

(ii) *To separate sin z into real and imaginary parts.*

We have $\sin z = \sin (x + iy)$

$= \sin x \cos (iy) + \cos x \sin (iy)$

[$\because \cos (iy) = \cosh y, \sin (iy) = i \sinh y$]

(iii) *To separate cos z into real and imaginary parts.*

We have $\cos z = \cos (x + iy)$

$= \cos x \cos (iy) - \sin x \sin (iy)$

$= \cos x \cosh y - i \sin x \sinh y$.

(iv) *To separate tan z into real and imaginary parts.*

We have $\tan z = \dfrac{\sin z}{\cos z} = \dfrac{\sin (x+iy)}{\cos(x+iy)}$

$$= \frac{\sin(x+iy)}{\cos(x+iy)} \times \frac{2\cos(x-iy)}{2\cos(x-iy)} = \frac{2\sin(x+iy)\cos(x-iy)}{2\cos(x+iy)\cos(x-iy)}$$

$$= \frac{\sin[(x+iy)+(x-iy)] + \sin[(x+iy)-(x-iy)]}{\cos[(x+iy)+(x-iy)] + \cos[(x+iy)-(x-iy)]}$$

$$= \frac{\sin 2x + \sin(2iy)}{\cos 2x + \cos(2iy)}$$

$$= \frac{\sin 2x + i\sinh 2y}{\cos 2x + \cosh 2y} \qquad [\because \sin(i\theta) = i\sinh\theta,\ \cos(i\theta) = \cosh\theta]$$

$$= \left(\frac{\sin 2x}{\cos 2x + \cosh 2y}\right) + i\left(\frac{\sinh 2y}{\cos 2x + \cosh 2y}\right).$$

(v) *To separate sinh z into real and imaginary parts.*

We have $\sinh z = \sinh(x + iy)$

$$= \frac{1}{i}\sin[i(x+iy)] \qquad [\because \sin(i\theta) = i\sinh\theta]$$

$$= \frac{1}{i^2}\sin(ix + i^2y)$$

$$= -i\sin(ix - y) \qquad [\because i^2 = -1]$$

$$= -i[\sin(ix)\cos y - \cos(ix)\sin y]$$

$$= -i[i\sinh x\cos y - \cosh x\sin y]$$

$$= -i^2\sinh x\cos y + i\cosh x\sin y$$

$$= \sinh x\cos y + i\cosh x\sin y.$$

(vi) *To separate tanh z into real and imaginary parts.*

We have $\tanh z = \dfrac{\sinh z}{\cosh z} = \dfrac{\sinh(x+iy)}{\cosh(x+iy)}$

$$= \frac{1}{i}\left[\frac{\sin i(x+iy)}{\cos i(x+iy)}\right] \qquad [\because \sin(i\theta) = i\sinh\theta,\ \cos(i\theta) = \cosh\theta]$$

$$= \frac{1}{i}\frac{\sin(ix-y)}{\cos(ix-y)} \qquad [\because i^2 = -1]$$

$$= -\frac{1}{i}\frac{\sin(y-ix)}{\cos(y-ix)} \qquad [\because \cos(-x) = \cos x,\ \sin(-x) = -\sin x]$$

$$= -\frac{i}{i^2}\,\frac{2\sin(y-ix)\cos(y+ix)}{2\cos(y-ix)\cos(y+ix)}$$ [Multiplying the Nr. and the Dr. b cos (y + ix) which is the conjugate complex of cos (y – ix)]

$$= i\,\frac{\sin[(y-ix)+(y+ix)]+\sin[(y-ix)-(y+ix)]}{\cos[(y-ix)+(y+ix)]+\cos[(y-ix)-(y+ix)]}$$

[∵ 2 sin A cos B = sin (A + B) + sin (A – B) and 2 cos A cos B = cos (A + B) + cos (A – B)

$$= i\,\frac{\sin 2y+\sin(-2ix)}{\cos 2y+\cos(-2ix)} = i\,\frac{\sin 2y-\sin(2ix)}{\cos 2y+\cos(2ix)}$$

$$= i\,\frac{\sin 2y - i\sinh 2x}{\cos 2y+\cosh 2x} = \frac{i\sin 2y - i^2\sinh 2x}{\cos 2y+\cosh 2x}$$

$$= \frac{\sinh 2x + i\sin 2y}{\cos 2y+\cosh 2x} = \frac{\sinh 2x}{\cosh 2x+\cos 2y} + i\,\frac{\sin 2y}{\cosh 2y+\cos 2y}$$

(vii) *To separate sech z into real and imaginary parts.*

We have $\operatorname{sech} z = \dfrac{1}{\cosh z} = \dfrac{1}{\cosh(x+iy)}$

$$= \frac{1}{\cos i(x+iy)} = \frac{1}{\cos(ix-y)} = \frac{1}{\cos(y-ix)} = \frac{2\cos(y+ix)}{2\cos(y+ix)\cos(y-ix)}$$

$$= 2.\frac{\cos y\cos(ix)-\sin y\sin(ix)}{\cos[(y+ix)+(y-ix)]+\cos[(y+ix)-(y-ix)]}$$

$$= 2.\frac{\cos y\cosh x - i\sin y\sinh x}{\cos 2y+\cos(2ix)} = \frac{2\cos y\cosh x}{\cosh 2x+\cos 2y} - i\,\frac{2\sin y\sinh x}{\cosh 2x+\cos 2y}.$$

SOLVED EXAMPLES

Example 1:

Show that exp (± i π/2) = ± i.

Solution:

Since exp (± iθ) = cos θ ± i sin θ, we have

$$\exp\left(\pm i\frac{\pi}{2}\right) = \cos\frac{\pi}{2} \pm i\sin\frac{\pi}{2} = 0 \pm i.1 = \pm i.$$

Example 2:

Prove that

$\{\sin)(\alpha - \theta) + e^{-\alpha i} \sin \theta\}^n = \sin^{n-1} \alpha \{\sin)(\alpha - n\theta) + e^{-\alpha i} \sin n\theta\}$

Solution:

L.H.S. $= \{\sin \alpha \cos \theta - \cos \alpha \sin \theta + (\cos \alpha - i \sin \alpha) \sin \theta\}^n$

$= \{\sin \alpha \cos \theta - i \sin \alpha \sin \theta\}^n = \sin^n \alpha (\cos \theta - i \sin \theta)^n$

$= \sin^n \alpha (\cos n\theta - i \sin n\theta)$, by De-Moivre's theorem

and R.H.S. $= \sin^{n-1} \alpha\alpha \{\sin (\alpha - n\theta) + e^{-\alpha i} \sin n\theta\}$

$= \sin^{n-1} \alpha \{\sin \alpha \cos n\theta - \cos \alpha \sin n\theta + (\cos \alpha - i \sin \alpha) \sin n\theta\}$

$= \sin^{n-1} \alpha \{\sin \alpha \cos n\theta - i \sin \alpha \sin n\theta\}$

$= \sin^{n-1} \alpha.\sin \alpha.(\cos n\theta - i \sin n\theta) = \sin^n \alpha (\cos n\theta - i \sin n\theta)$.

$\therefore$ L.H.S. = R.H.S.

Example 3:

Show that $\sinh (x + y) \cosh (x - y) = \frac{1}{2}(\sinh 2x + \sinh 2y)$.

Solution:

The L.H.S. $= \sinh (x + y) . \cosh(x - y)$

$$= \frac{1}{2}\left[e^{(x+y)} - e^{-(x+y)}\right].\frac{1}{2}\left[e^{(x-y)} + e^{-(x-y)}\right]$$

$$= \frac{1}{4}\left[e^{2x} + e^{2y} - e^{-2y} - e^{-2x}\right]$$

$$= \frac{1}{2}.\left[\frac{1}{2}\left(e^{2x} - e^{-2x}\right) + \frac{1}{2}\left(e^{2y} - e^{-2y}\right)\right]$$

$$= \frac{1}{2}\left[\sinh 2x + \sinh 2y\right] = \text{R.H.S.}$$

Example 4:

Show that $\cos(\alpha + i\beta) + i \sin (\alpha + i\beta) = e^{-\beta} (\cos \alpha + i \sin \alpha)$.

Solution:

The L.H.S. $= \cos \alpha \cos i\beta - \sin \alpha \sin i\beta + i \sin \alpha \cos i\beta + i \cos \alpha \sin i\beta$

$= \cos \alpha (\cos i\beta + i \sin i\beta) + \sin \alpha (\cos i\beta + i \sin i\beta)$

$= (\cos i\beta + i \sin i\beta) (\cos \alpha + i \sin \alpha)$

$= (\cosh \beta - \sinh \beta)(\cos \alpha + i \sin \alpha)$

$[\because \cos i\beta = \cosh \beta, \sin i\beta = i \sinh \beta]$

$= e^{-\beta} (\cos \alpha + i \sin \alpha) =$ R.H.S.

Example 5:

If $\cosh \alpha = \sec \theta$, *prove that* $\tanh^2 \frac{1}{2}\alpha = \tan^2 \frac{1}{2}\theta$.

Solution:

We have $\cosh \alpha = \sec \theta$.

$$\therefore \frac{\cosh \alpha}{1} = \frac{1}{\cos \theta}$$

Applying componendo and dividendo, we get

$$\frac{\cosh\alpha - 1}{\cosh\alpha + 1} = \frac{1-\cos\theta}{1+\cos\theta} \text{ or } \frac{2\sinh^2 \frac{1}{2}\alpha}{2\cosh^2 \frac{1}{2}\alpha} = \frac{2\sin^2 \frac{1}{2}\theta}{2\cos^2 \frac{1}{2}\theta}$$

or $\tanh^2 \frac{1}{2}\alpha = \tan^2 \frac{1}{2}\theta$.

Note: The componendo and dividendo is that if

$\frac{a}{b} = \frac{c}{d}$, then $\frac{a-b}{a+b} = \frac{c-d}{c+d}$.

Example 6:

If $\tan \theta = \tanh x \cot y$ *and* $\tan \phi = \tanh x$ *any, show that*

$$\frac{\sin 2\theta}{\sin 2\phi} = \frac{\cosh 2x + \cos 2y}{\cosh 2x - \cos 2y}$$ **(Rohilkhand, 1988)**

Solution:

$$\text{L.H.S.} = \frac{\sin 2\theta}{\sin 2\phi} = \frac{2\tan\theta/\left(1+\tan^2\theta\right)}{2\tan/\left(1+\tan^2\phi\right)}$$

$$= \frac{\tan\theta}{1+\tan^2\theta} \times \frac{1+\tan^2\phi}{\tan\phi} = \frac{\tan\theta}{\tan\phi} \cdot \frac{1+\tan^2\phi}{1+\tan^2\theta}$$

$$= \frac{\tanh x \cot y}{\tanh x \tan y} \times \frac{1+\tanh^2 x \tan^2 y}{1+\tanh^2 x \cot^2 y},$$ putting the given values of $\tan \theta$ and $\tan \phi$

$$= \frac{\cos^2 y}{\sin^2 y} \cdot \frac{\cosh^2 x \cos^2 y + \sinh^2 x \sin^2 y}{\cosh^2 x \cos^2 y} \cdot \frac{\cosh^2 x \sin^2 y}{\cosh^2 x \sin^2 y + \sinh^2 x \cosh^2 y}$$

$$= \frac{\cosh^2 x \cosh^2 x \sin^2 y}{\cosh^2 x \sin^2 y + \sinh^2 \cos^2 y}$$

$$= \frac{\left(2\cosh^2 x\right)\left(2\cos^2 y\right) + \left(2\sinh^2 x\right)\left(2\sin^2 y\right)}{\left(2\cosh^2 x\right)\left(2\sin^2 y\right) + \left(2\sinh^2 x\right)\left(2\cos^2 y\right)} \qquad \textbf{(Note)}$$

$$= \frac{(1+\cosh 2x)(1+\cos 2y) + (\cosh 2x - 1)(1-\cos 2y)}{(1+\cosh 2x)(1-\cos 2y) + (\cosh 2x - 1)(1+\cos 2y)}.$$

[$\because$ 2 $\cosh^2 x$ = 1 + cosh 2x; 2 $\sinh^2 x$ = cosh 2x – 1]

$$= \frac{2(\cosh 2x + \cos 2y)}{2(\cosh 2x - \cos 2y)} = \frac{\cosh 2x + \cos 2y}{\cosh 2x - \cos 2y} = \text{R.H.S.}$$

Example 7:

Separate $\dfrac{\cos(x+iy)}{(x+iy)+1}$ *into real and imaginary parts.* **(Meerut, 1996)**

Solution:

We have $\dfrac{\text{cis}(x+iy)}{(x+iy)+1} = \dfrac{\cos(x+iy)}{(x+1)+iy} = \dfrac{\cos(x+iy)\{(x+1)-iy\}}{\{(x+1)+iy\}\{(x+1)-iy\}}$

multiplying the Nr. and the Dr. by the conjugate complex of the Dr.

$$= \frac{[\cos x \cos(iy) - \sin x \sin(iy)][(x+) - iy]}{(x+1)^2 - i^2 y^2}$$

$$= \frac{(\cos x \cosh y - i \sin x \sinh y)[(x+1) - iy]}{(x+1)^2 + y^2}$$

$$= \frac{[(x+1)\cos x \cosh y - y \sin x \sinh y] - i[(x+1)\sin x \sinh y + y \cos x \cosh y]}{(x+1)^2 + y^2}$$

= a – ib, where

a = [(x + 1) cos x cosh y – y sin x sinh y]/$\{(x + 1)^2 + y^2\}$

and b = [(x + 1) sin x sinh y + y cos x cosh y]/$\{(x + 1)^2 + y^2\}$.

Example 8:

Resolve $e^{\sin(x + iy)}$ into real and imaginary parts. **(Garhwal, 1999)**

Solution:

We have $e^{\sin(x + iy)}$

$= e^{\sin x \cos (iy) + \cos x \sin (iy)}$

$= e^{\sin x \cosh y + i \cos x \sinh y}$

$= e^{\sin x \cosh y} . e^{i \cos x \sinh y}$

$= e^{\sin x \cosh y} [\cos(\cos x \sinh y) + i \sin (\cos x \sinh y)],$

$[\because e^{i\theta} \cos \theta + i \sin \theta]$

which is of the form a + ib

Example 9:

Resolve $\sin^2(x + iy)$ into real and imaginary parts.

Solution:

We have

$\sin^2 (x + y) = [2 \sin^2 (x + iy)]$

$= \frac{1}{2}[1 - \cos 2 (x + iy)]$ $\quad [\because 2 \sin^2 \theta = 1 - \cos 2\theta]$

$= \frac{1}{2}[1 - \cos (2x + 2iy)]$

$= \frac{1}{2}[1 - \{\cos 2x \cos (2iy) - \sin 2x \sin (2iy)\}]$

$= \frac{1}{2}[1 - \cos 2x \cosh 2y + i \sin 2x \sinh 2y]$

$= \frac{1}{2}(1 - \cos 2x \cosh 2y) + i \frac{1}{2} (\sin 2x \sinh 2y).$

Example 10:

If $E\left(\frac{x-a+iy}{x+a+iy}\right) = P + iQ$, find P and Q. **(Meerut, 1994)**

Solution:

Let $\quad \frac{x-a+iy}{x+a+iy} = p + iq.$

Then $$p+iq=\frac{\{(x-a)+iy\}\{(x+a)-iy\}}{\{(x+a)+iy\}\{(x+a)-iy\}},$$

multiplying the Nr. and Dr. by the conjugate complex of the Dr.

$$=\frac{(x^2-a^2)+y^2+i(yx+ya-yx+ya)}{(x+a)^2+y^2}=\frac{(x^2+y^2-a^2)+2iya}{(x+a)^2+y^2}.$$

Equating real and imaginary parts, we have

$$p=\frac{x^2+y^2-a^2}{(x+a)^2+y^2} \text{ and } q=\frac{2ay}{(x+a)^2+y^2}.$$

Again, $P+iQ=E\left(\frac{x-a+iy}{x+a+iy}\right)=e^{p+iq}$ **(Note)**

$= e^p.e^{iq} = e^p (\cos q + i \sin q)$.

Equating real and imaginary parts, we have

$P = e^p \cos q$ and $Q = e^p \sin q$,

where $p=\frac{x^2+y^2-a^2}{(x+a)^2+y^2}$ and $q=\frac{2ay}{(x+a)^2+y^2}$.

Example 11:

If $\sin(\theta + if) = \rho(\cos\alpha + i\sin\alpha)$, *prove that*

$r^2 = \frac{1}{2}[\cosh 2\phi - \cos 2\theta]$ *and* $\tan\alpha = \tanh\phi \cot\theta$. **(Meerut, 1991)**

Solution:

We have $\sin(\theta + i\phi) = \rho\cos\alpha + i\rho\sin\alpha$

or $\sin\theta\cosh\phi + i\cos\theta\sinh\phi = r\cos\alpha + ir\sin\alpha$

$\therefore \sin\theta\cosh\phi = \rho\cos\alpha$ ...(1)

and $\cos\theta\sinh\phi = \rho\sin\alpha$. ...(2)

Squaring (1) and (2) and adding, we get

$\rho^2 = \sin^2\theta\cosh^2\phi + \cos^2\theta\sinh^2\phi$.

$\therefore 2\rho^2 = 2\sin^2\theta\cosh^2\phi + 2\cos^2\theta\sinh^2\phi$

$= (1 - \cos 2\theta)\cosh^2\phi + (1 + \cos 2\theta)\sinh^2\phi$, changing to double angles

$\therefore (\cosh^2\phi + \sinh^2\phi) - \cos 2\theta(\cosh^2\phi - \sinh^2\phi)$

$= \cosh 2\phi - \cos 2\theta$. $[\because \cosh^2\phi + \sinh^2\phi = \cosh 2\phi]$

$\therefore\ \rho^2 = \dfrac{1}{2}(\cosh 2\phi - \cos 2\theta).$ **Proved.**

Again, dividing (2) by (1), we get

$\dfrac{\cos\theta \sinh\phi}{\sin\theta \cosh\phi} = \dfrac{\sin\alpha}{\cos\alpha}$ or $\tan\alpha = \tanh\phi \cot\theta.$ **Proved.**

Example 12:

If tan (θ + iφ) = tan α + i sec α, then prove that

$e^{2\phi} = \pm \cot \dfrac{1}{2}\alpha$ *and* $2\theta = n\pi + \dfrac{1}{2}\pi + a.$

(Meerut, 1988; Kanpur, 1998)

Solution:

We have $\tan(\theta + i\phi) = \tan\alpha + i \sec\alpha,$

so that $\tan(\theta - i\phi) = \tan\alpha - i\sec\alpha.$

Now $\tan 2\theta = \tan[(\theta + i\phi) + (\theta - i\phi)$

$$= \frac{\tan(\theta + i\phi) + \tan(\theta - i\phi)}{1 - \tan(\theta + i\phi)\tan(\theta - i\phi)}$$

$$= \frac{(\tan\alpha + i\sec\alpha) + (\tan\alpha - i\sec\alpha)}{1 - (\tan\alpha + i\sec\alpha)(\tan\alpha - i\sec\alpha)} = \frac{2\tan\alpha}{1 - (\tan^2\alpha + \sec^2\alpha)}$$

$$= \frac{2\tan\alpha}{-2\tan^2\alpha} = -\frac{1}{\tan\alpha} = -\cot\alpha.$$

Thus $\tan 2\theta = -\cot\alpha = \tan\left(\dfrac{1}{2}\pi + \alpha\right).$

$\therefore\ 2\theta = \dfrac{1}{2}\pi + \alpha.$ (principal value)

or $2\theta = n\pi + \dfrac{1}{2}\pi + \alpha$ (general value) **Proved.**

Again, $\tan 2i\phi = \tan[(\theta + i\phi) - (\theta - i\phi)]$

$$= \frac{\tan(\theta + i\phi) - \tan(\theta - i\phi)}{1 + \tan(\theta + i\phi)\tan(\theta - i\phi)}$$

$$= \frac{(\tan\alpha + i\sec\alpha) - (\tan\alpha - i\sec\alpha)}{1 + (\tan^2\alpha + \sec^2\alpha)} = \frac{2i\sec\alpha}{2\sec^2\alpha} = i\cos\alpha$$

$\therefore\ i\tanh 2\phi = i\cos\alpha$, or $\tanh 2\phi = \cos\alpha$

$$\text{or}\quad \frac{e^{2\phi}-e^{-2\phi}}{2e^{-2\phi}}=\frac{1+\cos\alpha}{1-\cos\alpha},\ \text{or}\quad e^{4\phi}=\frac{2\cos^2\frac{1}{2}\alpha}{2\sin^2\frac{1}{2}\alpha}=\cot^2\frac{1}{2}\alpha.$$

$\therefore\ e^{2\phi} = \pm\cot\frac{1}{2}\alpha.$ **Proved.**

Example 13:

If $\sin(\theta + i\phi) = \cos\alpha + i\sin\alpha$, *then prove that* $\cos^2\theta = \sinh^2\phi = \pm\sin\alpha$. **(Meerut, 1996)**

Solution:

We have $\sin(\theta + i\phi) = \cos\alpha + i\sin\alpha$

or $\sin\theta\cosh\phi + i\cos\theta\sinh\phi = \cos\alpha + i\sin\alpha$.

Equating real and imaginary parts, we have

$\sin\theta\cosh\phi = \cos\alpha$ ...(1)

$\cos\theta\sinh\phi = \sin\alpha$. ...(2)

First we shall eliminate ϕ between (1) and (2).

First we shall eliminate ϕ between (1) and (2).

We have $\cosh^2\phi - \sinh^2\phi = 1$.

$$\therefore\ \frac{\cos^2\alpha}{\sin^2\theta}-\frac{\sin^2\alpha}{\cos^2\theta}=1\ \text{or}\ \frac{\cos^2\alpha\cos^2\theta-\sin^2\alpha\sin^2\theta}{\sin^2\theta\cos^2\theta}=1$$

or $\cos^2\alpha\cos^2\theta - \sin^2\alpha\sin^2\theta = \sin^2\theta\cos^2\theta$

or $\cos^2\theta(1 - \sin^2\alpha) - \sin^2\alpha(1 - \cos^2\theta) = \cos^2\theta(1 - \cos^2\theta)$

[changing all the terms to $\cos\theta$ and $\sin\alpha$]

or $\cos^2\theta - \cos^2\theta\sin^2\alpha - \sin^2\alpha + \sin^2\alpha\cos^2\theta = \cos^2\theta - \cos^4\theta$

or $\cos^4\theta = \sin^2\alpha$.

$\therefore\ \cos^2\theta = \pm\sin\alpha$. **Proved.**

Now we shall eliminate θ between (1) and (2).

We have $\sin^2\theta + \cos^2\theta = 1$.

Therefore from (1) and (2), we have

$$\frac{\cos^2\alpha}{\cosh^2\phi}+\frac{\sin^2\alpha}{\sinh^2\phi}=1$$

or $\cos^2\alpha\alpha\sinh^2\phi + \cosh^2\phi\sin^2\alpha = \cosh^2\phi\sinh^2\phi$

or $\sinh^2 \phi\,(1 - \sin^2 \alpha) + \sin^2 \alpha\,(1 + \sinh^2 \phi) = (1 + \sinh^2 \phi) \sinh^2 \phi$

[changing all the terms to $\sinh \phi$ and $\sin \alpha$]

or $\sinh^2 \phi - \sinh^2 \phi \sin^2 \alpha + \sin^2 \alpha + \sinh^2 \phi \sin^2 \alpha = \sinh^2 \phi + \sinh^4 \phi$

or $\sinh^4 \phi = \sin^2 \alpha$.

$\therefore\ \sinh^2 \phi = \pm \sin \alpha$.

Thus $\cos^2 \theta = \sinh^2 \phi = \pm \sin \alpha$. **Proved.**

Example 14:

Split into real and imaginary parts $e^{i\theta}/1(1 - ke^{i\phi})$. **(Meerut, 1994)**

Solution:

We have $$\frac{e^{i\theta}}{1-ke^{i\phi}} = \frac{e^{i\theta}\left(1-ke^{-i\phi}\right)}{\left(1-ke^{i\phi}\right)\left(1-ke^{-i\phi}\right)}$$

multiplying the Nr. and Dr. by the conjugate complex of the Dr.

[Note that $1 - ke^{i\phi} = 1 - k(\cos \phi + i \sin \phi) = 1 - k \cos \phi - ik \sin \phi$ so that its conjugate complex is

$1 - k \cos \phi + ik \sin \phi$ *i.e.,* $1 - k(\cos \phi - i \sin \phi)$ *i.e.,* $1 - ke^{-i\phi}$]

$$= \frac{e^{i\theta} - ke^{i(\theta-\phi)}}{1-k\left(e^{i\phi}+e^{-i\phi}\right)+k^2e^{i\phi}.e^{-i\phi}}$$

$$= \frac{(\cos\theta + i \sin\theta) - k\left[\cos(\theta-\phi) + i \sin(\theta-\phi)\right]}{1-2k\cos\phi+k^2}$$

$$= \left(\frac{\cos\theta - k \cos(\theta-\phi)}{1-2k\cos\phi+k^2}\right) + i\left(\frac{\sin\theta - k\sin(\theta-\phi)}{1-2k\cos\phi+k^2}\right).$$

EXERCISES

1. Prove that $\sin(\alpha + n\theta) - e^{i\alpha} \sin n\theta = e^{-in\theta} \sin \alpha$.
2. Prove that $\{\sin(\alpha + \theta) - e^{i\alpha} \sin \theta\}^n = \sin^n a e^{-n\theta i}$.
3. Show that $\dfrac{1+\tanh x}{1-\tanh x} = \cosh 2x + \sinh 2x$.
4. Show that $\tanh(x+y) = \dfrac{\tanh x + \tanh y}{1+\tanh x \tanh y}$.
5. Show that:

(i) $\sinh x + \sinh y = 2 \sinh\frac{1}{2}(x + y) \cosh\frac{1}{2}(x - y)$

(ii) $\cosh x - \cosh y = 2\sinh\frac{1}{2}(x + y) \sinh\frac{1}{2}(x - y)$.

6. If $\alpha = \log \tan\left(\frac{\pi}{4}+\frac{\theta}{2}\right)$, prove that

(i) $\tan\frac{\alpha}{2} = \tan\frac{\theta}{2}$;

(ii) $\cos \theta \cosh \alpha = 1$;

(iii) $\sinh \alpha = \tan \theta$;

(iv) $\tanh \alpha = \sin \theta$;

(v) if $\alpha = \theta + a_3\theta^3 + a_5\theta^5 + ...$,

show that $\theta - \alpha - a_3\alpha^3 + a_5\alpha^5 - ...$

7. Resolve into real and imaginary parts:

(i) $\sin (\alpha + i\beta)$,

(ii) $\cos (\alpha + i\beta)$,

(iii) $\cot (\alpha + i\beta)$,

(iv) $\sec (\alpha + i\beta)$,

(v) $\text{cosec} (\alpha + i\beta)$,

(vi) $\cosh (\alpha + i\beta)$,

(vii) $\coth (\alpha + i\beta)$.

8. Resolve $e^{\cosh(x + iy)}$ into real and imaginary parts.

9. Resolve $e^{\sinh(x + iy)}$ into real and imaginary parts.

10. If $\sin (\alpha + i\beta) = x + iy$, prove that

(i) $x \text{ cosec}^2 \alpha - y^2 \sec^2 \alpha = 1$.

(ii) $x^2 \text{ sech}^2 \beta + y^2 \text{ cosech}^2 \beta = 1$. **(Meerut, 1988)**

11. If $\cos^{-1} (x + iy) = A + iB$, prove that

(i) $x^2\text{sech}^2 B + y^2 \text{ cosech}^2 B = 1$

(ii) $x^2 \sec^2 A - y^2 \text{ cosec}^2 A = 1$.

12. If $\tan (\theta + i\phi) = \sin (x + iy)$, then prove that

$\coth y \sinh 2\phi = \cot x \sin 2\theta$. **(Meerut, 1998)**

13. If $\tanh (\alpha + i\beta) = \sin (x + iy)$, prove that

$\sinh 2\alpha \text{ cosec } 2\beta = \tan x \coth y$.

14. If $\tan(\alpha + i\beta) = x + iy$, prove that
    (i) $x^2 + y^2 + 2x \cot 2\alpha = 1$;
    (ii) $x^2 + y^2 - 2y \coth 2\beta + 1 = 0$;
    (iii) $x \coth 2\alpha + y \coth 2\beta = 1$.
15. If $\tan(\theta + i\phi) = \cos\alpha + i \sin\alpha$, prove that
    $\theta = \frac{n\pi}{2} + \frac{\pi}{4}$ and $\phi = \frac{1}{2}\log\tan\left(\frac{\pi}{4} + \frac{\alpha}{2}\right)$. **(Meerut, 1985, 94)**
16. If $\tan(\alpha + i\beta) = i$, where α and β are real, then prove that α is indeterminate and β is infinite. **(Rohilkhand, 1998)**
17. If $\sin(\theta + i\phi) = \tan\alpha + i \sec\alpha$, then prove that
    $\cos 2\theta \cosh 2\phi = 3$. **(Meerut, 1994)**
18. If $\cos(x + iy) = \cos\alpha + i \sin\alpha$, prove that
    (i) $\cosh 2y + \cos 2x = 2$
    (ii) $\sin^4 x = \sin^2\alpha$
    (iii) $\sinh^4 y = \sin^2\alpha$.
19. If $\cos(\theta + i\phi) = r(\cos\alpha + i \sin\alpha)$, then prove that
    $\phi = \frac{1}{2}\log\frac{\sin(\theta - \alpha)}{\sin(\theta + \alpha)}$. **(Kanpur, 1998)**
20. If $\cosh u = \sec\theta$, show that
    $u = \log\tan\left(\frac{\pi}{4} + \frac{\theta}{2}\right)$ and $\tanh^2\frac{u}{2} = \tan^2\frac{\theta}{2}$.
21. If $A + iB = C\tan(x + iy)$, then prove that $\tan 2x = \frac{2CA}{C^2 - A^2 - B^2}$.
    (Meerut, 1986)
22. If $u + iv = \cot(x + iy)$, show that $v = -\frac{\sinh 2y}{\cosh 2y - \cos 2x}$.
23. If $\cos(u + iv) = x + iy$, prove that
    $(1 + x)^2 + y^2 = (\cosh v + \cos u)^2$
    and $(1 - x)^2 + y^2 = (\cosh v - \cos u)^2$, where x, y, u, v are all real.
24. If $\cosh(x + iy) = p + iq$, where x, y, p and q are real, show that
    $\frac{p^2}{\cos^2 y} - \frac{q^2}{\sin^2 y} = 1$. **(Agra, 1988)**

25. If sin (x + iy) sin (q + if) = 1, show that

(i) $\tanh^2 y \cosh^2 \phi = \cos^2 \theta$;

(ii) $\tanh^2 \phi \cosh^2 y = \cos^2 x$.

26. If $\cos(\theta + i\phi)\cos(\alpha + i\beta) = 1$, prove that **(Agra, 1994)**

$\tanh^2 \phi \cosh^2 \beta = \sin^2 \alpha$ and $\tanh^2 \beta \cosh^2 \phi = \sin^2 \theta$.

27. If $x = 2\cos\alpha\cosh\beta$ and $y = 2\sin\alpha\sinh\beta$, prove that:

(a) $\sec(\alpha + i\beta) + \sec(\alpha - i\beta) = \dfrac{4x}{x^2 + y^2}$,

(b) $\sec(\alpha + i\beta) - \sec(\alpha - i\beta) = \dfrac{4iy}{x^2 + y^2}$.

28. If α, β be the imaginary cube roots of unity, prove that

$$\alpha e^{ax} + \beta e^{bx} = -\, e^{-x/2}\left[\sqrt{3}\sin\frac{\sqrt{3}}{2}x + \cos\frac{\sqrt{3}}{2}x\right].$$

29. Show that $\dfrac{\sin(\alpha + i\beta)}{\sin(\alpha - i\beta)} + \dfrac{\sin(\alpha - i\beta)}{\sin(\alpha + i\beta)} = 2 \cdot \dfrac{\left(e^{2\beta} + e^{-2\beta}\right)\cos 2\alpha}{4\sin^2\alpha + \left(e^{\beta} - e^{-\beta}\right)^2}$.

30. Put $(\alpha + be^{i\theta})^n$ in the form A + iB.

3

Logarithms of Complex Numbers

3.1 LOGARITHMS IN THE SET OF REAL NUMBERS

We know that if a and x are two real numbers such that $e^x = a$, then x is called the logarithm of a to the base e and we write it as

$$x = \log_e a.$$

Since for all real numbers x, we have $e^x > 0$, therefore, in the case of a real variable, $\log_e a$ exists if and only if a is positive. Also if a is a positive real number, $\log_e a$ is a unique real number.

3.2 LOGARITHMS OF COMPLEX NUMBERS

Now we shall extend our definition of logarithmic function to the set of complex numbers.

Definition: Let z and w be two complex numbers such that $e^z = w$. Then z is called a *logarithm of w* to the base e and we write it as $z = \log_e w$ or simply as $z = \log w$, the base e remaining understood.

Thus by definition, $z = \log w$ if and only if $w = e^z$.

Since $e^z \neq 0$ for any complex number z, therefore log w does not exist if $w = 0$.

Logarithm of a Complex Number is a Many Valued Function

Let $\log_e w = z$. Then $e^z = w$.

If n is any integer, we have

$e^{2n\pi i} = \cos 2n\pi + i \sin 2n\pi = 1 + i0 = 1.$

$\therefore\ e^z = e^z.1 = e^z e^{2n\pi i} = e^{z + 2n\pi i},$

which means that if log w = z, then we also have

$$\log w = z + 2n\pi i.$$

Thus logarithm of a complex number is a many-valued function.

3.3 PRINCIPAL AND GENERAL VALUES OF LOGARITHM OF A NON-ZERO COMPLEX NUMBER

Let z = x + iy be a non-zero complex number. Suppose

$$\log_e z = \alpha + i\beta,$$

where α and β are real. Then

$$z = e^{\alpha + i\beta} = e^{\alpha}\, e^{i\beta} = e^{\alpha} (\cos \beta + i \sin \beta).$$

Since e^{α} is a positive real number, therefore

$$z = e^{\alpha} (\cos \beta + i \sin \beta)$$

is a representation of z in a modulus-argument form. We have

$$e^{\alpha} = |z| = \sqrt{(x^2 + y^2)} \text{ and } \beta = \arg z = \tan^{-1}\left(\frac{y}{x}\right).$$

The equation $e^{a} = |z|$ has a unique solution $\alpha = \log |z|$, the real logarithm of the positive number |z|.

$$\therefore \quad \log_e z = \log |z| + i \arg z$$

i.e., $$\log_e (x + iy) = \log\sqrt{(x^2 + y^2)} + i \tan^{-1}\left(\frac{y}{x}\right).$$

Now arg z is a many-valued function and consequently log z is a many-valued function. If β_0 is the principal value of arg z *i.e.,* the value of arg z lying between $-\pi$ and π, then $\log |z| + i\beta_0$ is called *the principal value of* log z. Again if β_0 is the principal value of arg z, then $2n\pi + \beta_0$ is the general value of arg z and

$$\log |z| + i\beta_0 + 2n\pi i$$

is called *the general value of log z. Thus every non-zero complex number h as infinitely many logarithms which differ from one another by an integral multiple of $2\pi i$.*

It is usual to denote the general value of $\log_e z$ by $\log_e z$, using the first letter L as the capital letter, and the principal value by $\log_e z$, using the first letter *l* as the small letter. Thus we have

$$\log_e z = \log_e z + 2n\pi i, \text{ where n is any integer.}$$

If in the general value, we put n = 0, we get the principal value.

Remember:

$$\log_e (x + iy) = \log\sqrt{(x^2+y^2)} + i\tan^{-1}\left(\frac{y}{x}\right),$$

where $\tan^{-1}\left(\frac{y}{x}\right)$ represents the principal value of arg (x + iy)

and $$\log_e (x + iy) = \log\sqrt{(x^2+y^2)} + i\tan^{-1}\left(\frac{y}{x}\right) + 2n\pi i,$$

where n is any integer.

3.4 PROPERTIES OF THE LOGARITHMIC FUNCTION

If u and v are two non-zero complex numbers, then

Log (uv) = Log u + Log v ...(1)

and Log (u/v) = Log u – Log V. ...(2)

Proof: Let Log u = z_1 and Log v = z_2.

Then $e^{z}{}_1 = u$ and $e^{z}{}_2 = v$.

$\therefore\ e^{z}{}_1 e^{z}{}_2 = uv$, or $e^{z_1+z_2} = uv$.

By the definition of logarithm, we have

Log uv = $z_1 + z_2$ = Log u + Log v, which proves the result (1).

The result (2) can be proved similarly.

The equality (1) does not mean that the principal value of log uv is equal to the sum of the principal values of log u and log v. Note that the sum of the principal arguments of u and v need not equal to some value of log uv and each value of log uv is equal to some value of log u + log v.

A similar remark holds for the equality (2) also.

Remark: If z be a non-zero complex number and n be a positive integer, the equality $\log z^n = n \log z$ may not be true even for the general values.

As an illustration, we have

$$\text{Log } i^3 = \text{Log }(-i) = \left(2n\pi - \frac{1}{2}\pi\right)i,$$

while $$3\,\text{Log } i = 3\left(2m\pi + \frac{1}{2}\pi\right)i.$$

Now it is not correct to say that

$$3\left(2m\pi+\frac{1}{2}\pi\right)=2n\pi-\frac{1}{2}\pi.$$

Now it is not correct to say that

$$3\left(2m\pi+\frac{1}{2}\pi\right)=2n\pi-\frac{1}{2}\pi.$$

For, the left hand side may be written as

$$2(3m+1)\pi-\frac{1}{2}\pi,$$

which shows that the solution set on the left is only a subset of the solution set on the right.

Thus Log $i^3 \neq 3$ Log i.

3.5 WORKING RULE TO EVALUATE Log (x + iy) *i.e.*, TO EXPRESS Log (x + iy) IN THE FORM A + iB

First we put x + iy in the modulus-argument form. So let

$x + iy = r(\cos\theta + i\sin\theta)$.

Then $x = r\cos\theta$, $y = r\sin\theta$.

$\therefore\ r\sqrt{(x^2+y^2)}$ and $\theta = \tan^{-1}\left(\frac{y}{x}\right)$.

Now $(x + iy) = \text{Log}(r\cos\theta + ir\sin\theta) = \text{Log}\, r(\cos\theta + i\sin\theta)$

$= \log[r\{\cos(2n\pi+\theta) + i\sin(2n\pi+\theta)\}] = \log r\, e^{i(2n\pi+\theta)}$

$= \log r + \log e^{i(2n\pi+\theta)} = \log r + i(2n\pi+\theta)$

$$=\log\sqrt{(x^2+y^2)}+i\left\{2n\pi+\tan^{-1}\left(\frac{y}{x}\right)\right\}$$

$$=\frac{1}{2}\log(x^2+y^2)+i\left\{2n\pi+\tan^{-1}\left(\frac{y}{x}\right)\right\}$$

$= A + iB$,

where $A=\frac{1}{2}\log(x^2+y^2)$,

$$B=2n\pi+\tan^{-1}\left(\frac{y}{x}\right).$$

If we put n = 0, we obtain the principal value of Log (x + iy) written as log (x + iy). Thus

$$\log(x+iy)=\frac{1}{2}\log(x^2+y^2)+i\tan^{-1}\left(\frac{y}{x}\right).$$

Putting –y for y on both sides, we get

$$\log(x-iy)=\frac{1}{2}\log(x^2+y^2)-i\tan^{-1}\left(\frac{y}{x}\right).$$

3.6 LOGARITHM OF A POSITIVE REAL NUMBER IN THE SET OF COMPLEX NUMBERS

Let x be a positive real number.

Let $x = x + i0 = r(\cos\theta + i\sin\theta)$

Then $x = r\cos\theta$, $0 = r\sin\theta$.

$\therefore$ $r = x$ and $\theta = 0$.

$\therefore$ $\log x = \log x + i0 + 2n\pi i = \log x + 2n\pi i$.

Obviously only $n = 0$ gives a real value of log x.

Thus in the set of complex numbers a positive real number has an infinite number of logarithms out of which only one is real.

3.7 LOGARITHM OF A NEGATIVE REAL NUMBER

Let x be a positive real number so that –x is a negative real number. We have to find log (–x).

Let $-x = -x + i0 = r(\cos\theta + i\sin\theta)$.

Then $-x = r\cos\theta$, $0 = r\sin\theta$.

$\therefore$ $r = x$, and $\theta = \pi$.

$\therefore$ $\log(-x) = \log x(\cos\pi + i\sin\pi) = \log x\{\cos(2n\pi + \pi) + i\sin(2n\pi + \pi)\}$

$= \log x\, e^{(2n\pi+\pi)i} = \log x + \log e^{(2n\pi+\pi)i}$

$= \log x + (2n\pi + 1)\pi i$, which is never real.

The principal value of log (–x) is $\log x + i\pi$.

Hence, the principal value of the logarithm of a negative real number is the logarithm of the corresponding positive number added with πi.

3.8 THE GENERAL EXPONENTIAL FUNCTION a^z

If a and z are any two complex numbers, the general exponential function a^z is defined as

$$a^z = e^{z \log a} = \exp(z \log a).$$

Since log a is a many-valued function, a^z is also a many-valued function. Thus,

$$a^z = e^{z \log a} = e^{z(\log a + 2n\pi i)} = \exp[z(\log a + 2n\pi i)].$$

This gives the *general value of a^z*. The *principal value* of a^z is obtained by putting n = 0. Hence the principal value of a^z

$$= e^{z \log a} = \exp(a \log a).$$

3.9 TO SEPARATE $(\alpha + i\beta)^{p + iq}$ INTO REAL AND IMAGINARY PARTS

By the definition of the general exponential function, we have

$$(a + ib)^{p + iq} = e^{(p + iq)\log(a + ib)} = \exp[(p + iq)\log(a + ib)]$$

$$= \exp\left[(p+iq)\left\{\frac{1}{2}\log\left(\alpha^2+\beta^2\right)+i\tan^{-1}\left(\frac{\beta}{\alpha}\right)+2n\pi i\right\}\right]$$

$$= \exp\left[\left\{\frac{1}{2}p\log\left(\alpha^2+\beta^2\right)-q\tan^{-1}\left(\frac{\beta}{\alpha}\right)-2nq\pi\right\}\right.$$

$$\left.+i\left\{\frac{1}{2}q\log\left(\alpha^2+\beta^2\right)+p\tan^{-1}\left(\frac{\beta}{\alpha}\right)+2np\pi\right\}\right]$$

$= \exp(A + iB)$, where $A = \frac{1}{2}p\log\left(\alpha^2+\beta^2\right)-q\left\{\tan^{-1}\left(\frac{\beta}{\alpha}\right)+2n\pi\right\}$

and $B = \frac{1}{2}q\log\left(\alpha^2+\beta^2\right)+p\left\{\tan^{-1}\left(\frac{\beta}{\alpha}\right)+2n\pi\right\}$

$$= e^{A + iB} = e^A e^{iB} = e^A(\cos B + i\sin B).$$

Therefore the real part is $e^A \cos B$ and the imaginary part is $e^A \sin B$.

The principal value is obtained by putting n = 0.

Note: In the following examples it is sometimes found convenient to write e^x as exp (x).

SOLVED EXAMPLES

Example 1:

Express log $(1 + i)^{(1 - i)}$ in the form A + iB.

Solution:

We have

$\log (1 + i)^{(1 - i)} = (1 - i) \log (1 + i)$

$$= (1-i)\left[\frac{1}{2}\log\left(1^2+1^2\right)+i\tan^{-1}1\right] = (1-i)\left[\frac{1}{2}\log 2 + i\frac{1}{4}\pi\right]$$

$$= \left(\frac{1}{2}\log 2 + \frac{1}{4}\pi\right) + i\left(\frac{1}{4}\pi + \frac{1}{2}\log 2\right),$$ which is of the form A + iB.

Example 2:

Find the general value of log (–i).

Solution:

We have $-i = \left(\cos\frac{1}{2}\pi - i\sin\frac{1}{2}\pi\right) = e^{-i\pi/2}$

so that $\log (-i) = \log e^{-i\pi/2} = -i\pi/2$, giving the principal value.

$$\therefore \log (-i) = \log (-i) + 2n\pi i = -\left(\frac{i\pi}{2}\right) + 2n\pi i = \frac{1}{2}(4n-1)\pi i.$$

Example 3:

Find the general value of log (–3). **(Meerut, 1993)**

Solution:

Let $-3 = -3 + i0 = r (\cos \theta + i \sin \theta)$,

so that $-3 = r \cos \theta$ and $0 = r \sin \theta$.

These give $r^2 = 9$ *i.e.*, $r = 3$. Putting $r = 3$, we get

$\cos \theta = -1$ and $\sin \theta = 0$, giving $\theta = \pi$.

$\therefore -3 = 3 (\cos \pi + i \sin \pi) = 3.e^{i\pi}$.

$\therefore \log (-3) = \log \{3.e^{i\pi}.e^{2n\pi i}\}$ $(\because e^{2n\pi i} = 1)$

$= \log 3 + \log e^{(2n\pi + \pi) i} = \log 3 + (2n + 1) \pi i$.

The principal value of log (–3) *i.e.*, log (–3) is obtained by putting n = 0 in the above result. Thus $\log (-3) = \log 3 + i\pi$.

Example 4:

Find the principal and general value of log (–1 + i). **(Kanpur, 1992)**

Solution:

Let $-1 + i = r(\cos\theta + i\sin\theta)$,

so that $r\cos\theta = -1$ and $r\sin\theta = 1$.

Squaring and adding, we have

$r^2 = 2$, *i.e.*, $r = \sqrt{2}$.

Now $\cos\theta = -\frac{1}{\sqrt{2}}$ and $\sin q = \frac{1}{\sqrt{2}}$,

so that $\theta = \frac{3}{4}\pi$.

$$\therefore\ -1 + i = \sqrt{2}\left(\cos\frac{3}{4}\pi + i\sin\frac{3}{4}\pi\right) = \sqrt{2e^{(3\pi/4)i}}$$

$\therefore$ the general value is

$$\log(-1 + i) = \log\left\{\sqrt{2e^{(3\pi/4)i}\, e^{2n\pi i}}\right\}$$

$$= \log\sqrt{2} + \frac{3}{4}\pi i + 2n\pi i = \frac{1}{2}\log 2 + \left(2n + \frac{3}{4}\right)\pi i.$$

Putting $n = 0$, the principal value is given by

$$\log(-1 + i) = \frac{1}{2}\log 2 + \frac{3}{4}\pi i.$$

Example 5:

If $\log\sin(x + iy) = a + ib$, *show that*

(i) $\alpha = \frac{1}{2}\log\frac{1}{2}(\cosh 2y - \cos 2x)$,

(ii) $2\cos 2x = e^{2y} + e^{-2y} - 4e^{2\alpha}$,

(iii) $b = \tan^{-1}(\cot x \tanh y)$,

(iv) $\cos(x - \beta) = e^{2y}\cos(x + \beta)$.

Solution:

$\because\ \log\sin(x + iy) = \alpha + i\beta$,

$\therefore\ \sin(x + iy) = e^{\alpha + i\beta} = e^{\alpha} e^{i\beta}$

or $\sin x \cos iy + \cos x \sin iy = e^{\alpha}(\cos\beta + i\sin\beta)$

or $\sin x \cosh y + i\cos x \sinh y = e^{\alpha}\cos\beta + i e^{\alpha}\sin\beta$.

Equating real and imaginary parts, we have

$$\sin x \cosh y = e^{\alpha} \cos \beta \qquad ...(1)$$

and $$\cos x \sinh y = e^{\alpha} \sin \beta. \qquad ...(2)$$

Squaring and adding (1) and (2), we get

$$\sin^2 x \cosh^2 y + \cos^2 x \sinh^2 y = e^{2\alpha} (\cos^{2\beta} + \sin^2\beta)$$

or $$\frac{1}{2}(1 - \cos 2x)\cosh^2 y + \frac{1}{2}(1 + \cos 2x) \sinh^2 y = e^{2\alpha}$$

or $$\frac{1}{2}\left(\cosh^2 y + \sinh^2 y\right) - \frac{1}{2}\cos 2x\left(\cosh^2 y - \sinh^2 y\right) = e^{2\alpha}$$

or $$\frac{1}{2}\cosh 2y - \frac{1}{2}\cos 2x = e^{2\alpha}$$

or $$\frac{1}{2}(\cosh 2y - \cos 2x) = e^{2\alpha}. \qquad ...(A)$$

$\therefore$ $$2\alpha = \log \frac{1}{2}(\cosh 2y - \cos 2x),$$

or $\alpha = \frac{1}{2}\log\frac{1}{2}(\cosh 2y - \cos 2x)$, which proves the result (i).

From the result (A), we have

$$\cosh 2y - \cos 2x = 2e^{2\alpha}$$

or $$\cosh 2y = \cos 2x + 2e^{2\alpha}$$

or $$2 \cosh 2y = 2 \cos 2x + 4e^{2\alpha}$$

or $2 \cos 2x = 2 \cosh 2y - 4e^{2\alpha} = e^{2y} + e^{-2y} - 4e^{2\alpha}$, which proves the result (ii).

Example 6:

Separate log sin (x + iy) into real and imaginary parts.

(Kanpur, 1996)

Solution:

We have

$\log \sin (x + iy) = \log (\sin x \cos iy + \cos x \sin iy)$

$= \log (\sin x \cosh y + i \cos x \sinh y)$

$$= \frac{1}{2}\log\left(\sin^2 x \cosh^2 y + \cos^2 x \sinh^2 y\right) + i \tan^{-1}\left(\frac{\cos x \sinh y}{\sin x \cosh y}\right) + 2n\pi i$$

$$= \frac{1}{2}\log\frac{1}{2}(1-\cos 2x)\cosh^2 y + \frac{1}{2}(1+\cos 2x)\sinh^2 y$$

$$+ i \tan^{-1} (\cot x \tanh y) + 2n\pi i$$

$$= \frac{1}{2}\log\left\{\frac{1}{2}\left(\cosh^2 y + \sinh^2 y\right) - \frac{1}{2}\cos 2x\left(\cosh^2 y - \sinh^2 y\right)\right\}$$

$$+ i \tan^{-1} (\cot x \tanh y) + 2n\pi i$$

$$= \frac{1}{2}\log\left(\frac{1}{2}\cosh 2y - \frac{1}{2}\cos 2x\right) + i\left[2n\pi + \tan^{-1}(\cot x \tanh y)\right]$$

$$= \frac{1}{2}\log\left[\frac{1}{2}(\cosh 2y - \cos 2x)\right] + i\left[2n\pi + \tan^{-1}(\cot x \tanh y)\right],$$

which is of the form P + iQ.

To find out the principal value, we put n = 0 in the above result. So we have

$$\log \sin (x + iy) = \frac{1}{2}\log\left[\frac{1}{2}(\cosh 2y - \cos 2x)\right] + i\tan^{-1}(\cot x \tanh y).$$

Example 7:

Prove that

$$\log\left(\frac{1}{1-e^{i\alpha}}\right) = \log\left(\frac{1}{2}\operatorname{cosec}\frac{\alpha}{2}\right) + i\left(\frac{\pi}{2} - \frac{\alpha}{2}\right).$$

(Meerut, 1991; Gorakhpur, 92; Garhwal, 91)

Solution:

We have

$$\log\frac{1}{1-e^{i\alpha}} = \log\frac{1}{1-\cos\alpha - i\sin\alpha} \qquad [\because e^{i\alpha} = \cos\alpha + i\sin\alpha]$$

$$= \log\frac{1}{2\sin^2\frac{1}{2}\alpha - 2i\sin\frac{1}{2}\alpha\cos\frac{1}{2}\alpha} = \log\frac{1}{2\sin\frac{1}{2}\alpha\left(\sin\frac{1}{2}\alpha - i\cos\frac{1}{2}\alpha\right)}$$

$$= \log\frac{1}{2\sin\frac{1}{2}\alpha\left\{\cos\left(\frac{1}{2}\pi - \frac{1}{2}\alpha\right) - i\sin\left(\frac{1}{2}\pi - \frac{1}{2}\alpha\right)\right\}}$$

$$= \log\frac{1}{2\sin\frac{1}{2}\alpha\, e^{-i(\pi/2-\alpha/2)}} \qquad [\because e^{-i\theta} = \cos\theta - i\sin\theta]$$

$$= \log\left[\left(\frac{1}{2}\operatorname{cosec}\frac{1}{2}\alpha\right).e^{i(\pi/2-\alpha/2)}\right] = \log\left(\frac{1}{2}\operatorname{cosec}\frac{1}{2}\alpha\right) + \log e^{i(\pi/2-\alpha/2)}$$

$$= \log\left(\frac{1}{2}\operatorname{cosec}\frac{1}{2}\alpha\right) + i\left(\frac{1}{2}\pi - \frac{1}{2}\alpha\right).$$

Example 8:

Prove that $i^i = e^{-(4n+1)\pi/2}$.

(Kanpur, 1990; Lucknow, 91; Gorakhpur 92)

Solution:

We have $i^i = e^{i \log i}$, by def. of a^z

$= \exp(i \log i) = \exp[i(\log i + 2n\pi i)]$. ...(1)

Now $\log i = \log\left(\cos\frac{1}{2}\pi + i\sin\frac{1}{2}\pi\right)$ $\left[\because i = \cos\frac{1}{2}\pi + i\sin\frac{1}{2}\pi\right]$

$\therefore \log e^{i\pi/2} = i\pi/2$.

Substituting for log i in (1), we have

$$i^i = \exp\left[i\left(i\frac{1}{2}\pi + 2n\pi i\right)\right] = \exp\left[i^2\left(2n + \frac{1}{2}\right)\pi\right]$$

$= \exp[-(4n+1)\pi/2] = e^{-(4n+1)\pi/2}$ ...(2)

Note 1: Putting n = 0 in (2), we get the principal value of $i^i = e^{-\pi/2}$.

Note 2: Putting n = 0, 1, 2, ... in (2), the various values of i^i are

$e^{-\pi/2}, e^{-5\pi/2}, e^{-9\pi/2}, e^{-13\pi/2}, \ldots$

which is a geometrical progression with common ratio $e^{-2\pi}$.

Example 9:

If $i^{\alpha + i\beta} = e^x(\cos y + i\sin y)$, *then prove that*

$$x = -\frac{1}{2}(4n+1)\pi\beta \text{ and } y = \frac{1}{2}(4n+1)\pi\alpha.$$

Solution:

We have

$e^x(\cos y + i\sin y) = i^{(\alpha + i\beta)}$. [given]

Now $i^{(\alpha + i\beta)} = \exp[(\alpha + i\beta)\log i]$

$= \exp[(a + i\beta)(\log i + 2n\pi i)]$

$= \exp [(\alpha + i\beta) (\log e^{i\pi/2} + 2n\pi i)]$ $\left[\because e^{i\pi/2} = \cos\frac{1}{2}\pi + i \sin\frac{1}{2}\pi = i\right]$

$= \exp\left[(\alpha + i\beta)\left(i\frac{1}{2}\pi + 2n\pi i\right)\right] = \exp[(i\alpha - \beta)(4n+1)\pi/2]$

$= e^{(i\alpha - \beta)(4n+1)\pi/2} = e^{-(4n+1)\pi\beta/2} e^{i(4n+1)\pi\alpha/2}$

$= e^{-(4n+1)\pi\beta/2} [\cos \{(4n + 1) \pi\alpha/2\} + i \sin \{(4n + 1) \pi\alpha/2\}]$.

$\therefore e^x (\cos y + i \sin y) = e^{-(4n+1)\pi\beta/2} [\cos \{(4m + 1) \pi\alpha/2\} +$
$i \sin \{(4n + 1\} \pi\alpha/2\}]$.

Equating real and imaginary parts, we get

$e^x \cos y = e^{-(4n+1) pb/2} \cos \{(4n + 1) \pi\alpha/2\}$ ...(1)

and $e^x \sin y = e^{-(4n+1)\pi\beta/2} \sin \{(4n + 1) \pi\alpha/2\}$. ...(2)

Squaring and adding (1) and (2), we get

$(e^x)^2 = [e^{-(4n+1)\pi\beta/2}]^2$

or $e^x = e^{-(4n+1)\pi\beta/2}$ or $x = -(4n + 1) \pi\beta/2$.

Dividing (2) by (1), we get

$\tan y = \tan \{(4n + 1) \pi\alpha/2\}$ or $y = (4m + 1) \pi\alpha/2$.

Example 10:

If $i^{i^{...ad inf.}} = A + iB$, principal values only being considered, prove that

(i) $\tan \frac{1}{2}\pi A = B/A$, *and*

(ii) $A^2 + B^2 = e^{-\pi\beta}$. **(Garhwal, 1983, 93; Gorakhpur, 91)**

Solution:

We have $i^{i^{...ad inf.}} = A + iB$.

$\therefore i^{A + iB} = A + iB$

or $\exp [(A + iB) \log i] = A + iB$, taking the principal value of $i^{A + iB}$

or $\exp [(A + iB) \log \left(\cos\frac{1}{2}\pi + i \sin\frac{1}{2}\pi\right) = A + iB$

or $\exp [(A + iB) \log e^{i\pi/2}] = A + iB$

or $\exp \left[(A + iB) i\frac{1}{2}\pi\right] = A + iB$

or $$\exp\left[-\frac{1}{2}\pi B+i\frac{1}{2}\pi A\right]=A+iB$$

or $$e^{-B\pi/2}\, e^{\pi Ai/2} = A + iB$$

or $$e^{-B\pi/2}\left(\cos\frac{1}{2}\pi A+i\sin\frac{1}{2}\pi A\right)=A+iB$$

Equating real and imaginary parts on both sides, we have

$$e^{-B\pi/2}\cos\frac{1}{2}\pi A=A, \qquad \text{...(1)}$$

and $$e^{-B\pi/2}\sin\frac{1}{2}\pi A=B.$$

Dividing (2) by (1), we have

$$\tan\frac{1}{2}\pi A=\frac{B}{A}.$$

Squaring and adding (1) and (2), we get

$$\left(e^{-B\pi/2}\right)^2\left(\cos^2\frac{1}{2}\pi A+\sin^2\frac{1}{2}\pi A\right)=A^2+B^2$$

or $$e^{-B\pi} = A^2 + B^2.$$

Example 11:

IF $p^{a+ib} = (x+iy)^{m+in}$ and the principal values are considered, prove that

(i) $a = \frac{1}{2} m \log_p (x^2 + y^2) - n \tan^{-1}\left(\frac{y}{x}\right) \log_p e,$

(ii) $\log_p (x^2 + y^2) = \frac{2(\alpha m+\beta n)}{m^2+n^2}.$ **(Meerut, 1984)**

Solution:

We have $p^{\alpha+i\beta} = (x+iy)^{m+in}$.

Taking logarithm of both sides, we have

$(\alpha + i\beta)\log_e p = (m + in)\log_e (x + iy)$

$$=(m+n)\left[\frac{1}{2}\log_e\left(x^2+y^2\right)+i\tan^{-1}\left(\frac{y}{x}\right)\right],$$

$$=\left[\frac{1}{2}m\log_e\left(x^2+y^2\right)-n\tan^{-1}\left(\frac{y}{x}\right)\right]$$

$$+\ i\left[\frac{1}{2}n\log_e\left(x^2+y^2\right)+m\tan^{-1}\left(\frac{y}{x}\right)\right]$$

Equating real and imaginary parts on both sides, we get

$$\left[\frac{1}{2}n\log_e\left(x^2+y^2\right)+m\tan^{-1}\left(\frac{y}{x}\right)\right] \quad ...(1)$$

$$\text{and } \beta\log_e p = \frac{1}{2}m\log_e\left(x^2+y^2\right)+m\tan^{-1}\left(\frac{y}{x}\right). \quad ...(2)$$

(i) From (1), we have

$$\alpha\log_e p = \frac{1}{2}m\log_e(x^2+y^2) - n\tan^{-1}\left(\frac{y}{x}\right)$$

$$\text{or } (\alpha\log_e p).(\log_p e) = \left[\frac{1}{2}m\log_e\left(x^2+y^2\right)-n\tan^{-1}\left(\frac{y}{x}\right)\right].\log_e e$$

multiplying both sides by $\log_p e$

$$\text{or } \alpha = \frac{1}{2}m\log_p\left(x^2+y^2\right)-n\tan^{-1}\left(\frac{y}{x}\right).\log_p e.$$

$[\because \log_b \alpha \times \log_a \beta = 1;\ \log_a m \times \log_b \alpha = \log_b m]$

$$\therefore\ \alpha = \frac{1}{2}m\log_p\left(x^2+y^2\right)-n\tan^{-1}\left(\frac{y}{x}\right)\log_p e.$$ **Proved.**

(ii) Multiplying (1) by m and (2) by n and adding, we get

$$(\alpha m+\beta n)\log_e p = \frac{1}{2}(m^2+n^2)\log_e(x^2+y^2)$$

$$\text{or } (\alpha m+\beta n)\log_e p\times\log_e = \frac{1}{2}(m^2+n^2)\log_e(x^2+y^2)\times\log_p e,$$

multiplying both sides by $\log_p e$

$$\text{or } (\alpha m+\beta n).1 = \frac{1}{2}(m^2+n^2)\log_p(x^2+y^2)$$

$$\text{or } \log_p(x^2+y^2) = \frac{2(\alpha m+\beta n)}{\left(m^2+n^2\right)}.$$

Example 12:

If $(a+ib)^p = m^{x+iy}$*, then prove that*

$$\frac{y}{x} = \frac{2\tan^{-1}\left(\frac{b}{a}\right)}{\log\left(a^2 + b^2\right)},$$

where only principal values are considered.

(Meerut, 1989; Garhwal, 90)

Solution:

Given that $(a + ib)^p = m^{(x + iy)}$.

Taking logarithm of both sides, we have

$$p \log (a + ib) = (x + iy) \log m$$

or $$p\left[\frac{1}{2}\log\left(a^2 + b^2\right) + i \tan^{-1}\left(\frac{b}{a}\right)\right] = x \log m + iy \log m.$$

Equating real and imaginary parts, we get

$$\frac{1}{2} p \log\left(a^2 + b^2\right) = x \log m \quad ...(1)$$

and $$p \tan^{-1}\left(\frac{b}{a}\right) = y \log m \quad ...(2)$$

Dividing (2) by (1), we get

$$\frac{y}{x} = \frac{p\tan^{-1}\left(\frac{b}{a}\right)}{\left(\frac{p}{2}\right)\log\left(a^2 + b^2\right)} = \frac{2\tan^{-1}\left(\frac{b}{a}\right)}{\log\left(a^2 + b^2\right)}.$$

Example 13:

Prove that if $(1 + i \tan a)^{1 + i \tan \beta}$ can have real values, one of them is $(\sec \alpha)^{\sec^2 \beta}$. **(Garhwal, 1993, 95; Meerut, 94)**

Solution:

We have $(1 + i \tan \alpha)^{1 + i \tan \beta}$

$= \exp [(1 + i \tan \beta) \log (1 + i \tan \alpha)]$, taking the principal value

$$= \exp\left[(1 + i \tan \beta)\left\{\log\sqrt{\left(1 + \tan^2 \alpha\right)} + i \tan^{-1}(\tan \alpha)\right\}\right]$$

$= \exp [(1 + i \tan \beta) (\log \sec \alpha + i\alpha)]$

= exp [log sec α – α tan β + i (α + tan β log sec α)]

= exp [log sec α – α tan β] · exp [i (α + tan β log sec α)]

= exp [log sec α – α tan β] · [cos {α + tan β log sec α} + i sin {α + tan β log sec α}].

Nos if this value is real, the imaginary part is equal to zero.

∴ sin {α + tan β log sec α} = 0

or α + tan β log sec α = 0

or α = – tan β log sec α. ...(1)

Also then $(1 + i \tan \alpha)^{1 + i \tan \beta}$ = exp [log sec α – α tan β] · (cos 0 + i sin 0)

= exp [log sec α + tan β · tan β log sec α], substituting for a from (1)

= exp [(1 + $\tan^2$ β) log sec α] = exp [$\sec^2$ β log sec α]

= exp [log $(\sec \alpha)^{\sec^2 \beta}$] = $(\sec \alpha)^{\sec^2 \beta}$.

This is one of the values since $(1 + i \tan \alpha)^{1 + i \tan \beta}$ can have infinite number of values.

Example 14:

Find the general value of $\log_4 (-2)$.

Solution:

We have $\log_4 (-2) = \dfrac{\log_3(-2)}{\log_e 4}$

$$= \frac{\log_e(-2) + 2n\pi i}{\log_e 4 + 2m\pi i}, \text{ where m and n are any integers}$$

$$= \frac{\log_e\{2(\cos\pi + i \sin\pi) + 2n\pi i\}}{\log_e 4 + 2m\pi i} = \frac{\log_e(2e^{i\pi}) + 2n\pi i}{\log_e 4 + 2m\pi i}$$

$$= \frac{\log_e 2 + i\pi + 2n\pi i}{\log_e 4 + 2m\pi i}$$

$$= \frac{\log_e 2 + i(2n\pi + \pi)}{2\log_e 2 + 2m\pi i} = \frac{[\log_e 2 + i(2n\pi + \pi)][2\log_e 2 - 2m\pi i]}{(2\log_e 2 + 2m\pi i)(2\log_e 2 - 2m\pi i)}$$

$$= \frac{2(\log_e 2)^2 + 2m\pi(2n\pi + \pi) + 2i\{(2n + 1)\pi \log_e 2 - m\pi \log_e 2\}}{4(\log_e 2)^2 + 4m^2\pi^2}$$

$$= \frac{(\log_e 2)^2 + (2n+1)m\pi^2}{2(\log_e 2)^2 + 2m^2\pi^2} + i\frac{(2n+1-m)\pi \log_e 2}{2(\log_e 2)^2 + 2m^2\pi^2},$$

which is required general value of $\log_4 (-2)$.

Example 15:

If sin log $(i^i) = a + ib$, find a and b. Hence find cos (log i^i).

Solution:

First, we have

$$i^i = \exp(i \log i) = \exp\left[i \log\left(\cos\frac{1}{2}\pi + i \sin\frac{1}{2}\pi\right)\right] = \exp [i \log e^{i\pi/2}]$$

$$= \exp\left[i\left(\frac{i\pi}{2}\right)\right] = \exp\left(-\frac{\pi}{2}\right) = e^{-\pi/2}.$$

$$\therefore \log i^i = \log^{-\pi/2} = -\frac{\pi}{2}.$$

$$\therefore \sin(\log i^i) = \sin\left(-\frac{\pi}{2}\right) = -1.$$

Hence $\sin(\log i^i) = a + ib$

$\Rightarrow -1 = a + ib.$

Equating real and imaginary parts, we get

$a = -1,\ b = 0.$

$\therefore \sin(\log i^i) = -1.$

Now $\cos(\log i^i) = \sqrt{\{1 - \sin^2(\log i^i)\}} = \sqrt{\{1-1\}} = 0.$

Example 16:

If log sin $(x + iy) = a + ib$, show that

(i) $\alpha = \frac{1}{2}\log\frac{1}{2}$ *(cosh 2y – cos 2x),*

(ii) $2 \cos 2x = e^{2y} + e^{-2y} - 4e^{2\alpha}$,

(iii) $b = \tan^{-1}(\cot x \tanh y)$,

(iv) $\cos(x - \beta) = e^{2y} \cos(x + \beta)$.

Solution:

$\because \log \sin (x + iy) = \alpha + i\beta,$

$\therefore \sin (x + iy) = e^{\alpha + i\beta} = e^{\alpha} e^{i\beta}$

or $\sin x \cos iy + \cos x \sin iy = e^{\alpha} (\cos \beta + i \sin \beta)$

or $\sin x \cosh y + i \cos x \sinh y = e^{\alpha} \cos \beta + i e^{\alpha} \sin \beta.$

Equating real and imaginary parts, we have

$$\sin x \cosh y = e^{\alpha} \cos \beta \quad ...(1)$$

and $$\cos x \sinh y = e^{\alpha} \sin \beta. \quad ...(2)$$

Squaring and adding (1) and (2), we get

$$\sin^2 x \cosh^2 y + \cos^2 x \sinh^2 y = e^{2\alpha} (\cos^{2\beta} + \sin^2 \beta)$$

or $$\frac{1}{2}(1-\cos 2x)\cosh^2 y + \frac{1}{2}(1+\cos 2x) \sinh^2 y = e^{2\alpha}$$

or $$\frac{1}{2}\left(\cosh^2 y + \sinh^2 y\right) - \frac{1}{2}\cos 2x\left(\cosh^2 y - \sinh^2 y\right) = e^{2\alpha}$$

or $$\frac{1}{2}\cosh 2y - \frac{1}{2}\cos 2x = e^{2\alpha}$$

or $$\frac{1}{2}(\cosh 2y - \cos 2x) = e^{2\alpha}. \quad ...(A)$$

$\therefore$ $$2\alpha = \log \frac{1}{2}(\cosh 2y - \cos 2x),$$

or $\alpha = \frac{1}{2}\log\frac{1}{2}(\cosh 2y - \cos 2x)$, which proves the result (i).

From the result (A), we have

$$\cosh 2y - \cos 2x = 2e^{2\alpha}$$

or $$\cosh 2y = \cos 2x + 2e^{2\alpha}$$

or $$2 \cosh 2y = 2 \cos 2x + 4e^{2\alpha}$$

or $2 \cos 2x = 2 \cosh 2y - 4e^{2\alpha} = e^{2y} + e^{-2y} - 4e^{2\alpha}$, which proves the result (ii).

EXERCISES

1. Prove that $\log (1 + i) = \frac{1}{2}\log 2 + i\left(2n\pi + \frac{1}{4}\pi\right)$. **(Jiwaji, 1982)**

2. Prove that $\log (-5) = \log 5 + (2n\pi + \pi) i$.

3. Find the general value of log i.
4. Find the general value of log (–i).
5. Find the general value of log $\sqrt{i}$.
6. Show that log $(1 + e^{i\theta}) = \log\left(2\cos\frac{1}{2}\theta\right) + \frac{1}{2}i\theta$, if $-\pi < \theta < \pi$.
7. Prove that $\log\left(\frac{a+ib}{a-ib}\right) = 2i\tan^{-1}\left(\frac{b}{a}\right)$.
8. Show that $i\log\frac{x-i}{x+i} = \pi - 2\tan^{-1}x$. **(Gorakhpur, 1991)**
9. Show that $\tan\left(i\log\frac{a-ib}{a+ib}\right) = \frac{2ab}{a^2-b^2}$.

(Meerut, 1989; Gorakhpur, 90)

10. Prove that log (1 + i tan α) = log sec α + αi.
11. Prove that sin (log i^i) = – 1.
12. If $u = \log\tan\left(\frac{\pi}{4}+\frac{\theta}{2}\right)$, prove that

   (i) $\tanh\frac{u}{2} = \tan\frac{\theta}{2}$.

   (ii) $\theta = -i\log\tan\left(\frac{\pi}{4}-\frac{iu}{2}\right)$.

13. Prove that log (1 + cos 2θ + i sin 2θ) = log (2 cos θ) + iθ, iφ – π < θ ≤ π.
14. Prove that log (iβ) = log |β| ± i $\frac{1}{2}\pi$, + or – sign being taken as β is +ive or –ive.
15. Separate into real and imaginary parts: log sin (x + iy).
16. Prove that

   $$\log\left[\frac{\sin(x+iy)}{\sin(x-iy)}\right] = 2i\tan^{-1}(\cot x\tanh y).$$

17. Prove that log cos (x + iy) = $\frac{1}{2}\log\frac{1}{2}$ (cosh 2y + cos 2x) – i $\tan^{-1}$ (tan x tanh y).

18. Prove that $\log\left\{\dfrac{\cos(x-iy)}{\cos(x+iy)}\right\} = 2i \tan^{-1} (\tan x \tanh y)$.

19. Show that one of the values of $\log\dfrac{(1+i)\left(1+i\sqrt{3}\right)}{\sqrt{3}+i}$ is $\dfrac{1}{2}\log 2+i\dfrac{5}{12}\pi$.

20. If $\tan \log (x + iy) = a + ib$, where $a^2 + b^2 \neq 1$,

prove that $\tan \{\log(x^2 + y^2)\} = \dfrac{2a}{1-a^2-b^2}$.

21. If $\log \log (x + iy) = p + iq$, prove that

$$y = x \tan [\tan q \log (x^2 + y^2)^{1/2}].$$

22. If $\log \log \log (\alpha + i\beta) = p + iq$, prove that

(i) $e^{e^p \cos q}.\sin\left(e^p \sin q\right) = \dfrac{1}{2}\text{loig}\left(\alpha^2+\beta^2\right)$,

(ii) $e^{e^p \cos q}.\sin\left(e^p \sin q\right) = \tan^{-1}\left(\dfrac{\beta}{\alpha}\right)$.

23. Prove that $\log\dfrac{(a-b)+i(a+b)}{(a+b)+i(a-b)} = i\left\{2n\pi+\tan^{-1}\dfrac{2ab}{a^2-b^2}\right\}$.

24. If $(a_1 + ib_1)(a_2 + ib_2) \ldots (a_n + ib_n) = A + iB$, prove that

$\tan^{-1}\left(\dfrac{b_1}{a_1}\right)+\tan^{-1}\left(\dfrac{b_2}{a_2}\right)+\ldots+\tan^{-1}\left(\dfrac{b_n}{a_n}\right) = \tan^{-1}\dfrac{B}{A}$,

and $\left(a_1^2+b_1^2\right)\left(a_2^2+b_2^2\right)\ldots\left(a_n^2+b_n^2\right) = A^2+B^2$.

25. If $(1 + i)(1 + 2i)(1 + 3i) \ldots (1 + ni) = A + iB$,

show that $2.5.10\ldots(1 + n^2) = (A^2 + B^2)$.

26. If $i^{\alpha + i\beta} = \alpha + i\beta$, show that $\alpha^2 + \beta^2 = e^{-(4n+1)\pi\beta}$.

(Meerut, 1991; Rohilkhand, 95)

27. Prove that $i^a = \cos\left[\left(2m+\dfrac{1}{2}\right)\pi a\right]+i\sin\left[\left(2m+\dfrac{1}{2}\right)\pi a\right]$.

28. If $i^{i^i} = \cos\theta + i\sin\theta$, prove that $\theta = \left(2m+\dfrac{1}{2}\right)\pi\exp\left[-\left(2n+\dfrac{1}{2}\right)\pi\right]$.

29. If $(i^i)^i = \cos\theta - i\sin\theta$, show that $\theta = \frac{1}{2}\pi(4n-1)$.

30. Show that the sum of the moduli of the values of $(1+i)^{1+i}$, which are less than unity is

$$\left(\frac{1}{\sqrt{2}}\right).e^{3\pi/4}.\text{cosech }\pi.$$

31. Show that the principal value of $(a+ib)^{p+iq}/(a-ib)^{p-iq}$ is $\cos 2(p\alpha + q\log r) + i\sin 2(p\alpha + q\log r)$,

where $r = \sqrt{(a^2+b^2)}$ and $\alpha = \tan^{-1}\left(\frac{b}{a}\right)$.

32. Prove that the real part of the principal value of

$(i)^{\log(1+i)}$ is $\exp\left(-\frac{\pi^2}{8}\right)\times\cos\left(\frac{1}{4}\pi\log 2\right)$.

33. Prove that $(x + ix\tan y)^{\log(x\sec y) - iy}$ is real, when only principal values are considered.

34. Prove that the principal value of $(a+ib)^{c+id}$ is wholly real or wholly imaginary according as

$$\frac{1}{2}d\log\left(a^2+b^2\right)+c\tan^{-1}\left(\frac{b}{a}\right)$$

is an even or an odd multiple of $\frac{1}{2}\pi$.

In case it is wholly real, prove that $(a+ib)^{c+id} = (a^2+b^2)^{(c^2+d^2)/2c}$.

35. Prove that the general value of $(1+i\tan\alpha)^{-i}$ is

$\exp(\alpha + 2m\pi).[\cos(\log\cos\alpha) + i\sin(\log\cos\alpha)]$.

(Meerut, 1988, 95; Gorakhpur, 81; Rohilkhand, 81)

36. Prove that $\log_i i = \frac{4m+1}{4n+1}$, where m and n are integers.

37. Prove that :

$$\log\left(\frac{a+ib-x}{a+ib+x}\right) = \frac{1}{2}\log\left\{\frac{(a-x)^2+b^2}{(a+x)^2+b^2}\right\} + i\tan^{-1}\left[\frac{b}{a-x} - \tan^{-1}\frac{b}{a+x}\right].$$

38. If $\dfrac{(1+i)^{p+iq}}{(1-i)^{p-iq}} = \alpha + i\beta$, prove that one value of $\tan^{-1}\left(\dfrac{\beta}{\alpha}\right)$ is $\left(\dfrac{\pi p}{2}\right) + q\log 2$.

39. If $[\cos(\theta - i\phi)]^{x+iy} = A + iB$ and principal values are taken into consideration, then

$$\tan^{-1}\left(\frac{B}{A}\right) = \frac{1}{2}\, y \log(\cosh^2\phi - \sin^2\theta) + x\tan^{-1}(\tan\theta\tanh\phi).$$

40. If $\left\{\dfrac{a+x+iy}{a-x-iy}\right\}^{\lambda+\mu i} = X + iY$, prove that one of the values of $\tan^{-1}\left(\dfrac{Y}{X}\right)$ is $\lambda\tan^{-1}\left\{\dfrac{2ay}{a^2-x^2-y^2}\right\} + \dfrac{\mu}{2}\log\left\{\dfrac{(a+x)^2+y^2}{(a-x)^2+y^2}\right\}$.

4

Inverse Circular and Hyperbolic Functions of Complex Numbers

4.1 INVERSE CIRCULAR FUNCTIONS

Definitions: *The equation sin $\theta = x$ means that θ is the angle whose sine is x. To express θ explicitly in terms of x, a convenient notation 'sin ^{1} x' (read as sine inverse x) is introduced. Thus, $\theta = sin^{-1};)$: means that θ is the angle whose sine is x. Similarly 'cos^{-1} x' expresses an angle whose cosine isx, tan^x denotes an angle whose tangent is x, and so on.*

The quantities $\sin^{-1}$ x, $\cos^{-1}$ x, $\tan^{-1}$ x etc., are called *Inverse Circular Functions*. Sometimes $\sin^{-1}$ x is written as '*arc sine x*' with similar notations for other inverse functions.

Note: $\sin^{-1}$ x should not be confused with $(\sin x)^{-1}$ as

$$(\sin x)^{-1} = 1/\sin x.$$

4.2 GENERAL AND PRINCIPAL VALUES OF INVERSE CIRCULAR FUNCTIONS

Consider the equation $\theta = \sin^{-1} x$, where $-1 \leq x \leq 1$. In the set of real numbers, there are infinite values of θ which satisfy this equation. All these values of θ taken together form what we call the general value of $\sin^{-1}$x. Among the values of θ satisfying the equation $\theta = \sin^{-1}$x, the value which is numerically the smallest one is called the *principal value* of $\sin^{-1}$ x. If θ is the principal value of $\sin^{-1}$ x, obviously we must have $-\frac{1}{2}\pi \leq \theta \leq \frac{1}{2}\pi$.

If x is 0, the principal value of $\sin^{-1}$x is 0; if x is negative, the principal value of $\sin^{-1}x$ lies between $-\frac{1}{2}\pi$ It and 0; and if x is positive, the principal value of $\sin^{-1}x$ lies between 0 and $\frac{1}{2}\pi$. Thus, the principal value of $\sin^{-1}(-\frac{1}{\sqrt{2}})$ is $1/4\pi$ while the principal value of $\sin^{-1}(-\frac{1}{\sqrt{2}})$ is $-\frac{1}{4}\pi$.

If θ is the principal value of $\sin^{-1}x$, then all the angles given by $n\pi + (-1)^n\theta$, where n is any integer, have their sines equal to x. We call $n\pi + (-1)^n\theta$ as the general value of $\sin^{-1}x$ and denote it by $\sin^{-1}x$. Generally, if the first letter of an inverse circular function is small, we consider the principal value, while if the first letter is capital, it means the general value.

For example, $\sin^{-1}\frac{1}{2} = \frac{1}{6}\pi$, while $\sin^{-1}\frac{1}{2} = n\pi + (-1)^n\frac{1}{6}\pi$, where n is any integer.

Thus, the inverse sine of x is a many-valued function. Its principal value is denoted by $\sin^{-1}x$ and its general value is denoted by $\sin^{-1}x$. Also we have

$$\sin^{=1}x = np + (-l)^n \sin^{-1}x,$$

where n is any integer, positive or negative or zero.

Similarly, we may define the concepts of the principal and general values of the other inverse circular functions. The relations between the general and principal values of various inverse circular functions are as follows:

$$\mathbf{\sin^{-1}x = n\pi + (-l)^n \sin^{-1}x,}$$

$$\mathbf{\cos^{-1}x = 2n\pi \pm \cos^{-1}x,}$$

$$\mathbf{\tan^{-1}x = n\pi + (-l)^n \tan^{-1}x,}$$

$$\mathbf{\cos^{-1}x = 2n\pi \pm \cos^{-1}x,}$$

$$\mathbf{\sec^{-1}x = 2n\pi \pm \sec^{-1}x,}$$

$$\mathbf{\cot^{-1}x = n\pi + \cot^{-1}x,}$$

where n is any integer, positive or negative or zero.

The principal value of any inverse circular function is the smallest numerical value of that inverse circular function. It may be positive or negative. In case there are two values, one positive and the other negative, which are numerically equal and smallest, the principal value is taken as the positive one. For example, the principal value of $\cos^{-1}$ is $\frac{1}{3}\pi$, as out of the two numerically smallest values $\frac{1}{3}\pi$ and $-\frac{1}{3}\pi$ of $\cos^{-1}\frac{1}{2}\pi$ the value $\frac{1}{3}\pi$ is positive.

It is to be noted that by the value of an inverse circular function we usually mean its priocipal value unless the contrary is stated. If x is positive, the principal values of $\sin^{-1} x$, $\text{cosec}^{-1} x$ $\cos^{-1} x$, $\sec^{-1} x$, $\tan^{-1} x$ and $\cot^{-1} x$ all lie between 0 and $\frac{1}{2}\pi$. But if x is negative the principal values of $\sin^{-1} x$, $\text{cosec}^{-1} x$, $\tan^{-1} x$ and $\cot^{-1} x$ lie between $-\frac{1}{2}\pi$ and 0, while those of $\cos^{-1} x$ and $\sec^{-1}$ lie between $\frac{1}{2}\pi$ and π. Thus, the principal values of $\sin^{-1} x$ and $\tan^{-1}x$ (and therefore of $\text{cosec}^{-1} x$ and $\cot^{-1} x$) lie between $\frac{1}{2}\pi$ and $\frac{1}{2}\pi$, while those of $\cos^{-1} x$ and $\sec^{-1} x$ lie between 0 and π.

4.3 RELATIONS BETWEEN INVERSE FUNCTIONS

(a) *Principle of reciprocity:* We have

$$\sin^{-1} x = \text{cosec}^{-1} (1/x);$$

$$\cos^{-1} x = \sec^{-1} (1/x);$$

and $$\tan^{-1} x = \cot^{-1} (1/x).$$

(b) From the definition of an inverse circular function, it is clear that

$$\theta = \sin^{-1} (\sin \theta),$$

and $$x = \sin (\sin^{-1} x).$$

Similarly $\theta = \cos^{-1} (\cos \theta)$

and $$x = \cos (\cos^{-1} x),$$

$$\theta = \tan^{-1} (\tan \theta),$$

and $$x = \tan (\tan^{-1} x), \text{ etc.}$$

$$\sin^{-1} x = \cos^{-1} \sqrt{(1 - x^2)} = \tan^{-1}\left\{\frac{x}{\sqrt{(1-x)^2}}\right\}$$

$$\cos— x = \sin^{-1} \sqrt{(1 - x^2)} = \tan^{-1} \frac{\sqrt{(1-x^2)}}{x} \text{ etc.,}$$

and $$\tan^{-1} x = \sin^{-1} \left\{\frac{x}{\sqrt{(1-x)^2}}\right\} = \cos^{-1} \left\{\frac{1}{\sqrt{(1-x)^2}}\right\} \text{ etc.}$$

(c) We have,

(i) $\sin^{-1} (-x) = - \sin^{-1} x$;

(ii) $\cos^{-1}(-x) = \pi - \cos^{-1} x$;

and (iii) $\tan^{-1}(-x) = -\tan^{-1} x$.

4.4 SOME IMPORTANT RESULTS ABOUT INVERSE FUNCTIONS

(a) Complementary inverse functions: We have

(i) $\sin^{-1} x + \cos^{-1} x = \pi/2$;

(ii) $\tan^{-1} x + \cot^{-1} x = \pi/2$;

and (iii) $\sec^{-1} x + \operatorname{cosec}^{-1} x = \pi/2$.

(b) Important formulae: We have,

(i) $\tan^{-1} x + \tan^{-1} y = \tan^{-1}\{(x + y)/(1 - xy)\}$

(ii) $\tan^{-1} x - \tan^{-1} y = \tan\text{-}1\{(x - y)/(1 + xy)\}$;

(iii) $2 \tan^{-1} x = \tan^{-1}\{2x/(1 - x^2)\}$;

and (iv) $\tan^{-1} x + \tan^{-1} y + \tan^{-1} z = \tan^{-1} = \dfrac{x+y+z-xyz}{1-yz-zx-xy}$

Remark: The formula

$$\tan^{-1} x + \tan^{-1} y = \tan\text{–}^{1} \{(x + y)/(1 - xy)\}$$

does not mean that the sum of the principal values of $\tan^{-1}x$ and $\tan^{-1}y$ will necessarily be equal to the principal value of $\tan^{-1} \{(x + y)/(1 - xy)\}$.

This sum may be equal to the principal value of $\tan^{-1} \{(x + y)/(1 - xy)\}$ or it may be equal to some other value of $\tan^{-1} \{(r + y)/(1 - xy)\}$.

(c) Some more formulae: We have

(i) $\text{col}^{-1} x + \cot^{-1} y = \cot^{-1} \dfrac{xy-1}{x+y}$

(ii) $\cot^{-1} x - \cot^{-1} y = \cot^{-1} \dfrac{xy+1}{y-x}$

4.5 INVERSE CIRCULAR FUNCTIONS OF COMPLEX NUMBERS

If $\cos(x + iy) = u + iv$, then $x + iy$ is called an *inverse cosine* of $u + iv$ and is denoted by $\cos^{-1}(u + iv)$. Thus

if $(\cos x + iy) = u + iv$, then $\cos^{-1}(u + iv) = x + iy$.

Also, if $\cos(x + iy) = u + iv$, then

$u + iv = \cos\{2n\pi \pm (x + iy)\}$, where n is any integer.

Therefore by the above definition the *general value* of inverse cosine of $u + iv$ is

$2n\pi \pm (x + iy)$,

and is denoted by $\cos^{-1}(u + iv)$ *i.e.*, by writing the first letter 'C' as capital.

It follows that the inverse cosine of u + iv is a many-valued function. Its *principal value* is that value of $2n\pi \pm (x + iy)$ in which the real part lies between 0 and π. It is denoted by $\cos^{-1}(u + iv)$ *i.e.*, by writing the first letter 'c' as small. Thus

$$\cos^{-1}(u + iv) = 2n\pi \pm (x + iy) = 2n\pi \pm \cos^{-1}(u + iv).$$

In a like manner if $\sin(x + iy) = u + iv$, then $x + iy$ is called an inverse sine of $u + iv$ and is denoted by $\sin^{-1}(u + iv)$. Thus

if $\sin(x + iy) = u + iv$, then $\sin^{-1}(u + iv) = x + iy$.

Also, if $\sin(x + iy) = u + iv$, then

$u + iv = \sin(n\pi + (-1)^n (x + iy)\}$, where n is any integer.

Therefore the general value of inverse sine of $u + iv$ is

$n\pi + (-1)^n (x + iy)$,

and is denoted by $\sin^{-1}(u + iv)$, the first letter 'S' being written capital.

Therefore the inverse sine of $u + iv$ is a many-valued function. Its principal value is that value of $n\pi + (-1)^n (x + iy)$ for which the real part lies between $-\frac{1}{2}\pi$ and $\frac{1}{2}\pi$. It is denoted by $\sin^{-1}(u + iv)$, the first letter 's' being written small. Thus

$$\sin^{-1}(u + iv) = n\pi + (-1)^n \sin^{-1}(u + iv).$$

The other inverse circular functions may be defined similarly. For example if

$$\tan(x + iy) = u + iv,$$

then $\tan^{-1}(u + iv) = n\pi + (x + iy)$

$= np + \tan^{-1}(x + iy)$,

the principal value being that for which the real part lies between $-\frac{1}{2}\pi$ and $\frac{1}{2}\pi$.

The relations concecting the general and principal values of the remaining inverse circular functions may be written as follows:

$$\sec^{-1}(u + iv) = 2n\pi \pm \sec^{-1}(u + iv),$$

$\text{cosec}^{-1}(u + iv) = n\pi\ (-1)^n\ \text{cosec}^{-1}(u + iv)$,

and $\cot^{-1}(u + iv) = np + \cot^{-1}(u + iv)$.

It should be noted that the principal value for the sin, cosec, tan and cot is that value for which the real part lies between $-\frac{1}{2}p$ and $\frac{1}{2}p$, while for the case of and sec it is that value for which its real part lies between 0 and π.

4.6 INVERSE HYPERBOLIC FUNCTIONS

Let z and w be two complex numbers. If sinh w = z, then w is called the inverse hyperbolic sine of z and is written as

$w = \sinh^{-1} z$.

The other inverse hyperbolic functions $\cosh^{-1} z$, $\tanh^{-1} z$, $\text{cosech}^{-1}z$, $\text{sech}^{-1} z$ and $\coth^{-1}$ are defined similarly.

The values of inverse hyperbolic functions can also be expressed in terms of the logarithmic functions as shown below.

(i) To prove that $\sinh^{-1} z = \log\left[z+\sqrt{(z^2+1)}\right]$.

Proof: Let $\sinh^{-1} z = w$. The

$\sinh w = z.$ ...(1)

$\therefore\ \cosh w = \sqrt{(1+\sinh^2 w)} = \sqrt{(1+z^2)}.$ ...(2)

Adding (1) and (2), we get

$$\sinh w + \cosh w = z+\sqrt{(1+z^2)}$$

i.e., $\frac{1}{2}(e^w - e^{-w}) + \frac{1}{2}(e^w + e^{-w}) = z+\sqrt{(1+z^2)}$

i.e., $e^w = z+\sqrt{(1+z^2)}\ e^w = z+\sqrt{(1+z^2)}\ e^w = z+\sqrt{(1+z^2)}$.

$$\therefore\ w = \log\left[z+\sqrt{(1+z^2)}\right].$$

Hence $\sinh^{-1}z = \log\left[z+\sqrt{(z^2+1)}\right]\left[z+\sqrt{(z^2+1)}\right]$.

(ii) To prove that $\cosh^{-1} z = \log\left[z + \sqrt{(z^2 - 1)}\right]$.

Proof: Let $\cosh^{-1}z = w$. Then

$\cosh w = z.$ ...(1)

$$\therefore \sinh w = \sqrt{(\cosh^2 w - 1)} = \sqrt{(z^2 - 1)}$$

Adding (1) and (2), we get

$$\cosh w + \sinh w = z + \sqrt{(z^2 - 1)}$$

$$\text{or } e^w = z + \sqrt{(z^2 - 1)}$$

$$\therefore w = \log\left[z + \sqrt{(z^2 - 1)}\right].$$

$$\text{Hence } \cosh^{-1}z = \log\left[z + \sqrt{(z^2 - 1)}\right].$$

(iii) To prove that

$$\tanh^{-1} z = \frac{1}{2}\log\frac{1+z}{1-z}.$$

Proof: Let $\tanh^{-1}z = w$. Then $\tanh w = z$

$$\text{or } \frac{e^w - e^{-w}}{e^w + e^{-w}} = \frac{z}{1}.$$

Applying componendo and dividendo, we have

$$\frac{(e^w + e^{-w}) + (e^w - e^{-w})}{(e^w + e^{-w}) - (e^w - e^{-w})} = \frac{1+z}{1-z}.$$

$$\text{or } \frac{2e^w}{2e^{-w}} = \frac{1+z}{1-z} \text{ or } e^{2w} = \frac{1+z}{1-z}.$$

$$\therefore 2w = \log\frac{1+z}{1-z} \text{ or } w = \frac{1}{2}\log\frac{1+z}{1-z}.$$

Hence $\tanh^{-1}z = \frac{1}{2}\log\frac{1+z}{1-z}$.

(iv) Similarly, we can prove that

$$\coth^{-1} z = \frac{1}{2}\log\frac{z+1}{z-1}.$$

4.7 RELATIONS BETWEEN INVERSE HYPERBOLIC FUNCTIONS AND INVERSE CIRCULAR FUNCTIONS

(i) To prove that $\sinh^{-1}x = -i\sin^{-1}(ix)$.

Proof: Let $\sinh^{-1} x = y$.

Then $x = \sinh y$.

$\therefore\ ix = i\sinh y = \sin(iy)$.

$\therefore\ iy = \sin^{-1}(ix)$

or $y = \left(\frac{1}{i}\right)\sin^{-1}(ix) = -i\sin^{-1}(ix)$

Hence $\sinh^{-1}x = -i\sin^{-1}(ix)$.

(ii) To prove that $\cosh^{-1}x = -i\cos^{-1}(x)$.

Proof: Let $\cosh^{-1}x = y$. Then $x = \cosh y = \cos(iy)$.

$\therefore\ iy = \cos^{-1}x$

or $y = \left(\frac{1}{i}\right)\cos^{-1}x = -i\cos^{-1}x$.

Hence $\cosh^{-1}x = -i\cos^{-1}x$.

(iii) To prove that $\tanh^{-1}x = -i\tan^{-1}(ix)$.

Proof: Let $\sinh^{-1}x = y$.

Then $x = \tanh y$.

$\therefore\ ix = i\tanh y = \tan(iy)$

$\therefore\ iy = \tan^{-1}(ix)$

or·$y = \left(\frac{1}{i}\right)\tan^{-1}(ix) = -i\tan^{-1}(ix)$

Hence $\tanh^{-1}x = -i\tan^{-1}(ix)$.

SOLVED EXAMPLES

Example 1:

Express $\cos^{-1}(x + iy)$ in the form $A + iB$.

Solution:

Let $\cos^{-1}(x + iy) = A + iB$.

Then $\cos(A + iB) = x + iy$

or $\cos A \cos(iB) - \sin A \sin(iB) = x + iy$

or $\cos A \cosh B - \sin A \sinh B = x + iy$.

Equating real and imaginary parts on both sides, we get

$\cos A \cosh B = x,$ ...(1)

and $\sin A \sinh B = -y.$ ...(2)

Let us first eliminate B between (1) and (2). From (1) and (2), we have

$$\cosh B = \frac{x}{\cos A} \text{ and } \sinh B = -\frac{y}{\sin A}.$$

$$\therefore \frac{x^2}{\cos^2 A} - \frac{y^2}{\sin^2 A} = \cos^2 B - \sinh^2 B = 1$$

or $x^2\sin^2 - y^2\cos^2 A = \cos^2 A \sin^2 A$

or $x^2 \sin^2 A - y^2 (1 - \sin^2 A) = (1 - \sin^2 A)\sin^2 A$

or $x^2 \sin^2 A - y^2 + y^2 \sin^2 A = \sin^2 A - \sin^4 A$

or $\sin^4 A + (x^2 + y^2 - 1)\sin^2 A - y^2 = 0,$

which is a quadratic in $\sin^2 A$.

$$\therefore \sin^2 A = \frac{-\left(x^2 + y^2 - 1\right) \pm \sqrt{\left[\left(x^2 + y^2 - 1\right)^2 + 4y^2\right]}}{2}.$$

Since $\sin^2 A$ must be positive, therefore neglecting the –ive sign, we have

$$\sin^2 A = \frac{\sqrt{\left\{\left(x^2 + y^2 - 1\right)^2 + 4y^2\right\}} - \left(x^2 + y^2 - 1\right)}{2}$$

$$\therefore \quad \sin A = \pm\left[\frac{\sqrt{\left\{\left(x^2+y^2-1\right)^2+4y^2\right\}}-\left(x^2+y^2-1\right)}{2}\right]^{1/2}$$

or $$A = \pm\sin^{-1}\left[\frac{\sqrt{\left\{\left(x^2+y^2-1\right)^2+4y^2\right\}}-\left(x^2+y^2-1\right)}{2}\right]^{1/2} \qquad ...(3)$$

Now let us eliminate A between (1) and (2). From (1) and (2), we have

$\cos A = \dfrac{x}{\cosh B}$ and $\sin A = -\dfrac{y}{\sinh B}$.

$$\therefore \quad \frac{x^2}{\cosh^2 B} + \frac{y^2}{\sinh^2 B} = \cos^2 A + \sin^2 A = 1$$

or $x^2 \sinh^2 B + y^2 \cosh^2 B = \cosh^2 B \sinh^2 B$

or $x^2 \sinh^2 B = y^2(1 + \sinh^2 B) = (1 + \sinh^2 B) \sinh^2 B$

or $x^2 \sinh^2 B + y^2 + y^2 \sinh^2 B = \sinh^2 B + \sinh^4 B$

or $\sinh^4 B + (1 - x^2 - y^2) \sinh^2 B - y^2 = 0$,

which is a quadratic in $\sinh^2 B$.

$$\therefore \quad \sinh^2 B = \frac{-\left(1-x^2-y^2\right) \pm \sqrt{\left\{\left(1-x^2-y^2\right)^2+4y^2\right\}}}{2}.$$

But $\sinh^2 B$ is positive, so neglecting the –ive sign, we have

$$\sinh^2 B = \frac{\sqrt{\left\{\left(1-x^2-y^2\right)^2+4y^2\right\}}-\left(1-x^2-y^2\right)}{2}.$$

$$\therefore \quad \sinh B = \pm\left[\frac{\sqrt{\left\{\left(1-x^2-y^2\right)^2+4y^2\right\}}-\left(1-x^2-y^2\right)}{2}\right]^{1/2}$$

$$\text{or } B = \pm\sin^{-1}\left[\frac{\sqrt{\left\{\left(1-x^2-y^2\right)^2+4y^2\right\}-\left(1-x^2-y^2\right)}}{2}\right]^{1/2} \qquad ...(4)$$

Remark: The general value of $\cos^{-1}(x + iy)$ *i.e.*, $\cos^{-1}(x + iy)$ is given by

$\cos^{-1}(x + iy) = 2np \pm \cos^{-1}(x + iy) = 2np \pm (A + iB)$,

where A and B are as found in (3) and (4).

Example 2:

Express $\sin^{-1}(x + iy)$ in the form $A + iB$.

Solution:

Proceed exactly as in Ex. 1.

If we have already found $\cos^{-1}(x + iy)$, then $\sin^{-1}(x + iy)$ can also be deduced from it as shown below.

We have $\sin^{-1}(x + iy) = \frac{1}{2}\pi - \cos^{-1}(x + iy)$

$$= \frac{1}{2}\pi \pm \sin^{-1}\left[\frac{\sqrt{\left\{\left(x^2+y^2-1\right)^2+4y^2\right\}-\left(x^2+y^2-1\right)}}{2}\right]^{1/2}$$

$$\pm i\sinh^{-1}\left[\frac{\sqrt{\left\{\left(1-x^2-y^2\right)^2+4y^2\right\}-\left(1-x^2-y^2\right)}}{2}\right]^{1/2},$$

substituting for $\cos^{-1}(x + iy)$ from Ex. 1.

Example 3:

Express $\cosh^{-1}(x + iy)$ in the form $\alpha + i\beta$.

Solution:

Let $\cosh^{-1}(x + iy) = \alpha + i\beta$.

Then $x + iy = \cosh(\beta + i\beta) = \cos\{i(\alpha + i\beta)\} = \cos(i\alpha - \beta)$

$= \cos i\alpha \cos\beta + \sin i\alpha \sin\beta = \cosh\alpha \cos\beta + i \sinh\alpha \sin\beta.$

Equating real and imaginary parts on both sides, we have

$\cosh\alpha \cos\beta = x,$ ...(1)

and $\sinh\alpha \sin\beta = y.$ ...(2)

To find α, we eliminate β between (1) and (2). From eqns. (1) and (2), we have

$$\cos\beta = \frac{x}{\cosh\alpha} \text{ and } \sin\beta = \frac{y}{\sinh\alpha}.$$

$$\therefore \quad \frac{x^2}{\cosh^2\alpha} + \frac{y^2}{\sinh^2\alpha} = \cos^2\beta + \sin^2\beta = 1$$

or $x^2 \sinh^2\alpha + y^2 \cosh^2\alpha = \cosh^2\alpha \sinh^2\alpha.$

Proceeding as in Ex. 1, we get

$$\alpha = \pm \sinh^{-1}\left[\frac{\sqrt{\left\{\left(1 - x^2 - y^2\right)^2 + 4y^2\right\}} - \left(1 - x^2 - y^2\right)}{2}\right]^{1/2}.$$

Again to find β, we eliminate α between (1) and (2). From (1) and (2), we have

$$\cos\alpha = \frac{x}{\cos\beta} \text{ and } \sinh\alpha = \frac{y}{\sin\beta}.$$

$$\therefore \quad \frac{x^2}{\cos^2\beta} - \frac{y^2}{\sin^2\beta} = \cosh^2\alpha - \sinh^2\alpha = 1.$$

Proceeding as in Ex. 1, we get

$$\beta = \pm \sin^{-1}\left[\frac{\sqrt{\left\{\left(x^2 + y^2 - 1\right)^2 + 4y^2\right\}} - \left(x^2 + y^2 - 1\right)}{2}\right]^{1/2}.$$

Example 4:

Explain the meaning of $\sin^{-1}x$, *when* x *is real and greater than 1, and find the value of* $\sin^{-1} 2$.

Solution:

We know that if α is a real number, then sin α lies between -1 and 1. Therefore if x is a real number greater than 1, then $\sin^{-1} x$ is not a real number, but is a complex number.

We have $\sin^{-1} 2 = n\pi + (-1)^n \sin^{-1} 2$.

Let $\sin^{-1} 2 = u + iv$, where u and v are real.

Then $\sin (u + iv) = 2$

or $\sin u \cos iv + \cos u \sin iv = 2$ or $\sin u \cosh v + i \cos u \sinh v = 2$.

Equating real and imaginary parts, we get

$\sin u \cosh v = 2,$ ...(1)

and $\cos u \sinh v = 0.$ ...(2)

From (2), $\cos u \sinh v = 0$. But $\sinh v = 0$ gives $v = 0$ which makes $\sin^{-1} 2$ real. So we have $\sinh v \neq 0$.

$\therefore \cos u = 0$, or $u = \pm \frac{1}{2}\pi$.

Putting $u = \pm\frac{1}{2}\pi$ in (1), we get

$\cosh v = \pm 2$.

$\therefore v = \cosh^{-1} (\pm 2) = \log \left(\sqrt{3} \pm 2\right) \left[\because \cosh^{-1} x = \log\left\{x + \sqrt{(x^2 - 1)}\right\}\right]$

$\therefore \sin^{-1} 2 = u + iv = \pm\frac{1}{2}\pi + i \log\left(\sqrt{3} \pm 2\right)$

$\therefore \sin^{-1} 2 = n\pi + (-1)^n \left[\pm\frac{1}{2}\pi + i \log\left(\sqrt{3} \pm 2\right)\right].$

Example 5:

Express $\tan^{-1} (x + iy)$ as the sum of real and imaginary parts.

Solution:

Let $\tan^{-1} (x + iy) = A + iB$. Then

$\tan (A + iB) = x + iy,$

and $\tan (A - iB) = x - iy$, by eq. ıting complex conjugates.

Now $\tan 2A = \tan [(A + iB) + (A - iB)]$

$$= \frac{\tan(A + iB) + \tan(A - iB)}{1 - \tan(A + iB).\tan(A - iB)} = \frac{(x + iy) + (x - iy)}{1 - (x + iy)(x - iy)}$$

$$= \frac{2x}{1-(x^2+y^2)} = \frac{2x}{1-x^2+y^2}.$$

$$\therefore \quad 2A = \tan^{-1}\frac{2x}{1-x^2-y^2}$$

or $$A = \frac{1}{2}\tan^{-1}\frac{2x}{1-x^2-y^2}.$$

Again $\tan(2iB) = \tan[(A+iB)-(A-iB)]$

$$= \frac{\tan(A+iB)-\tan(A-iB)}{1+\tan(A+iB)\tan(A-B)} = \frac{(x+iy)-(x-iy)}{1+(x+iy)(x-iy)} = \frac{2iy}{1+x^2+y^2}.$$

$$\therefore \; i\tanh 2B = \frac{2iy}{1+x^2+y^2} \qquad [\because \tan(i\theta) = i\tanh\theta]$$

or $\tanh 2B = \dfrac{2y}{1+x^2+y^2}$ or $2B = \tanh^{-1}\dfrac{2y}{1+x^2+y^2}$

or $B = \dfrac{1}{2}\tanh^{-1}\dfrac{2y}{1+x^2+y^2}.$

Hence $$\tan^{-1}(x+iy) = A+iB = \frac{1}{2}\tan^{-1}\frac{2x}{1-x^2-y^2} + i\frac{1}{2}\tanh^{-1}\frac{2y}{1+x^2+y^2}.$$

Note: If we are to find the general value $\tan^{-1}(x+iy)$, then

$$\tan^{-1}(x+iy) = n\pi + \tan^{-1}(x+iy)$$

$$= n\pi + \frac{1}{2}\tan^{-1}\frac{2x}{1-x^2-y^2} + i\frac{1}{2}\tanh^{-1}\frac{2y}{1+x^2+y^2}.$$

Example 6:

Prove that $\sinh^{-1}(\cot x) = \log(\cot x + \operatorname{cosec} x)$.

Solution:

Let $\sinh^{-1}(\cot x) = y$.

Then $\sinh y = \cot x$. ...(1)

$$\therefore \cosh y = \sqrt{(1+\sinh^2 y)} = \sqrt{(1+\cot^2 x)} = \operatorname{cosec} x. \quad ...(2)$$

Adding (1) and (2), we have

$$\sinh y + \cosh y = \cot x + \operatorname{cosec} x$$

or $e^y = \cot x + \operatorname{cosec} x$

or $\quad y = \log(\cot x + \operatorname{cosec} x)$.

$\therefore \quad \sinh^{-1}(\cot x) = \log(\cot x + \operatorname{cosec} x)$.

Example 7:

Express $\tanh^{-1}(x + iy)$ *into the form* $\alpha + i\beta$ *and hence deduce the value of* $\tanh^{-1}(iy)$.

Solution:

Let $\tanh^{-1}(x + iy) = \alpha + i\beta$

so that $x + iy = \tanh(\alpha + i\beta) = \left(\frac{1}{i}\right) \tan[i(\alpha + i\beta)]$.

$[\because \tan i\theta = i \tanh \theta]$

$\therefore\ i(x + iy) = \tan(i\alpha - \beta)$ or $ix - y = \tan\{-(\beta - i\alpha)\}$

or $-(y - ix) = -\tan(\beta - i\alpha)$ $\quad [\because \tan(-z) = -\tan z]$

or $\tan(\beta - i\alpha) = y - ix$. ...(1)

Equating the complex conjugates of both sides of (1), we have

$\tan(\beta + i\alpha) = y + ix$. ...(2)

Now $\tan 2\beta = \tan[(\beta - i\alpha) + (\beta + i\alpha)]$

$$= \frac{\tan(\beta - i\alpha) + \tan(\beta + i\alpha)}{1 - \tan(\beta - i\alpha)\tan(\beta + i\alpha)} = \frac{(y - ix) + (y + ix)}{1 - (y - ix)(y + ix)}, \text{ from (1) and (2)}$$

$$= \frac{2y}{1 - (y^2 + x^2)} = \frac{2y}{1 - x^2 - y^2}.$$

$$\therefore\ 2\beta = \tan^{-1}\frac{2y}{1 - x^2 - y^2}, \text{ or } \beta = \frac{1}{2}\tan^{-1}\left(\frac{2y}{1 - x^2 - y^2}\right) \qquad ...(3)$$

Again, $\tan(2i\alpha) = \tan[(\beta + i\alpha) - (\beta - i\alpha)]$

$$= \frac{\tan(\beta + i\alpha) - \tan(\beta - i\alpha)}{1 + \tan(\beta + i\alpha)\tan(\beta - i\alpha)} = \frac{(y + ix) - (y - ix)}{1 + (y + ix)(y - ix)} = \frac{2ix}{1 + y^2 + x^2}.$$

$$\therefore\ i\tanh 2\alpha = \frac{2ix}{1 + x^2 + y^2} \text{ or } \tanh 2\alpha = \frac{2x}{1 + x^2 + y^2}$$

$$\text{or } \alpha = \frac{1}{2}\tanh^{-1}\left(\frac{2x}{1 + x^2 + y^2}\right). \qquad ...(4)$$

$\therefore\ \tanh^{-1}(x + iy) = \alpha + i\beta$

$$= \frac{1}{2}\tanh^{-1}\left(\frac{2x}{1-x^2+y^2}\right) + i\frac{1}{2}\tan^{-1}\frac{2y}{\left(1-x^2-y^2\right)} \text{ [From (3) and (4) ...(5)}$$

To deduce the value of $\tanh^{-1}$ (iy), put x = 0 on both sides of (5).

Then $\tanh^{-1}(iy) = \frac{1}{2}\tanh^{-1}0 + i\frac{1}{2}\tan^{-1}\frac{2y}{1-y^2}$

$$= 0 + i\frac{1}{2}.2\tan^{-1}y \qquad \left[\because 2\tan^{-1}y = \tan^{-1}\frac{2y}{1-y^2}\right]$$

$= i \tan^{-1} y.$

$\therefore \tanh^{-1}(iy) = i \tan^{-1} y.$

Example 8:

If $\sin^{-1}(\theta + i\phi) = \alpha + i\beta$, then prove that $\sin^2\alpha$ and $\cosh^2\beta$ are the roots of the equation

$x^2 - x(1 + \theta^2 + \phi^2) + \theta^2 = 0.$

Solution:

We have $\sin^{-1}(\theta + i\phi) = \alpha + i\beta$

so that θ + if = $\sin(\alpha + i\beta)$ or θ + if = $\sin\alpha \cosh\alpha + i\cos\alpha \sinh\beta$.

Equating real and imaginary parts, we have

$\sin\alpha \cosh\beta = \theta,$ ...(1)

and $\cos\alpha \sinh\beta = \phi.$ ...(2)

Now $1 + \theta^2 + \phi^2 = 1 + \sin^2\alpha \cosh^2\beta + \cos^2\alpha \sinh^2\beta$, from eqns. (1) and (2)

$= 1 + \sin^2\alpha \cosh^2\beta + (1 - \sin^2\alpha)(\cosh^2\beta - 1)$

$= 1 + \sin^2\alpha \cosh^2\beta + \cosh^2\beta - 1 - \sin^2\alpha \cosh^2\beta + \sin^2\alpha$

$= \sin^2\alpha + \cosh^2\beta.$

Thus $\sin^2\alpha + \cosh^2\beta = 1 + \theta^2 + \phi^2.$

Also from (1), $\sin^2\alpha \cosh^2\beta = \theta^2.$

$\therefore$ $\sin^2\alpha$ and $\cosh^2\beta$ are the roots of the equation

$x^2 - (\sin^2\alpha + \cosh^2\beta)x + \sin^2\alpha \cosh^2\beta = 0$

or $x^2 (1 + \theta^2 + \phi^2) x + \theta^2 = 0.$

EXERCISES

1. Show that $\sin^{-1}(\cos\theta + i\sin\theta) = \cos^{-1}\sqrt{(\sin\theta)} + i\log\left[\sqrt{(\sin\theta)} + \sqrt{(1+\sin\theta)}\right]$, where θ is a positive acute angle.

(Meerut, 1985P)

2. Prove that one of the values of $\sin^{-1}(\cos\theta + i\sin\theta)$ is $\cos^{-1}\sqrt{(\sin\theta)} + i\log\left[\sqrt{(\sin\theta)} + \sqrt{(1+\sin\theta)}\right]$.

3. Show that $\cos^{-1}(\cos\theta + i\sin\theta) = \sin^{-1}\sqrt{(\sin\theta)} + i\log\left\{\sqrt{(1+\sin\theta)} - \sqrt{(\sin\theta)}\right\}$, where θ is a positive acute angle.

4. Separate $\cos^{-1}(\cos\theta + i\sin\theta)$ into real and imaginary parts.

5. If $\cos^{-1}(\alpha + i\beta) = \theta + if$, then prove that:

   (i) $\alpha^2\operatorname{sech}^2\phi + \beta^2\operatorname{cosech}^2\phi = 1$,

   (ii) $\alpha^2\sec^2\theta - \beta^2\operatorname{cosec}^2\theta = 1$.

6. If $\sin^{-1}(x + iy) = \tan^{-1}(u + iv)$, show that

$$[(x-1)^2 + y^2]\,[(x+1)^2 + y^2] = \frac{\left(x^2+y^2\right)^2}{\left(u^2+v^2\right)^2}.$$

7. Prove that $\sin^{-1}(\operatorname{cosec}\theta) = \{2n + (-1)^n\}\frac{1}{2}\pi + i(-1)^n\log\cot\frac{1}{2}\theta$.

(Garhwal, 1993, 94)

8. Show that $\sin^{-1}(ix) = n\pi + i(-1)^n\log\left\{x + \sqrt{\left(1+x^2\right)}\right\}$.

9. Prove that $\tan^{-1}(\cos\theta + i\sin\theta) = n\pi + \frac{1}{4}\pi - \frac{1}{2}i\log\tan\left(\frac{1}{4}\pi - \frac{1}{2}\theta\right)$.

(Garhwal 1985; Agra 89; Rohilkhand 90, 91)

10. Prove that $\tan^{-1}(\cos q + i\sin q) = n\pi + \frac{1}{4}\pi + \frac{1}{2}i\log\tan\left(\frac{1}{4}\pi + \frac{1}{2}\theta\right)$.

(Meerut 1984S)

11. If $x > y$, then show that $\tan^{-1}\left(\frac{x+iy}{x-iy}\right) = \frac{\pi}{4} + \frac{i}{2}\log\frac{x+y}{x-y}$.

12. Prove that $\tan^{-1}\left[i\dfrac{x-a}{x+a}\right]=-\dfrac{1}{2}i\log\left(\dfrac{a}{x}\right)$. **(Meerut, 1982)**

13. If $\cosh x = \sec\theta$, then prove that $x = \log(\sec\theta \pm \tan\theta)$.

14. Show that $\sinh^{-1} x = \tanh^{-1}\dfrac{x}{\sqrt{(1+x^2)}}$.

15. Prove that $\tanh^{-1}x = \sinh^{-1}\left\{\dfrac{x}{\sqrt{(1-x^2)}}\right\}$.

16. Prove that $\coth^{-1}\left(\dfrac{2}{x}\right)=\sinh^{-1}\left\{\dfrac{x}{\sqrt{(4-x^2)}}\right\}$.

17. Prove that $\tan^{-1}\left(\dfrac{\tan 2\theta+\tanh 2\phi}{\tan 2\theta-\tanh 2\phi}\right)+\tan^{-1}\left(\dfrac{\tan\theta-\tanh\phi}{\tan\theta+\tanh\phi}\right) = \tan^{-1}$ $(\cot\theta\coth\phi)$.

18. If $\cos^{-1}(u + iv) = \alpha + i\beta$, prove that $\cos^2\alpha$ and $\cosh^2\beta$ are the roots of the equation
$x^2 - (1 + u^2 + v^2)x + u^2 = 0$.

19. If $\cosh^{-1}(x + iy) + \cosh^{-1}(x - iy) = \cosh^{-1} a$,
show that $2(a - 1)x^2 + 2(a + 1)y^2 = a^2 - 1$. **(Meerut, 1983)**

5

Expansion of Some Trigonometrical Functions

5.1 INTRODUCTION

In this chapter, we shall expand some trigonometrical functions like $\cos^n \Theta$, $\sin^n \Theta$, $\cos^n \Theta \sin^m \Theta$, where n, m are positive integers. The following result will be used in this chapter.

Let $\quad x = \cos \Theta + \sin \Theta$

Then $\quad \frac{1}{x} = \frac{1}{\cos \Theta + i \sin \Theta}$

$= (\cos \Theta + i \sin \Theta)^{-1} = \cos \Theta - i \sin \theta$ [by De Moivre's Theorems]

$\therefore \quad x + \frac{1}{x} = 2 \cos \Theta$

and $\quad x - \frac{1}{x} = 2 \sin \Theta$

By De Moivre'ss Theorem, we have

$$x^n = \cos n \theta + i \sin n \Theta$$

and $\quad \frac{1}{x^n} = \cos n \theta - i \sin m \Theta$

$\therefore \quad x^n + \frac{1}{x^n} = 2 \cos m \Theta$

and $\quad x^n - \frac{1}{x^n} = 2 i \sin n \Theta.$

5.2 EXPANSION OF COS^N Θ IN A SERIES OF COSINES OF MULTIPLE OF Θ, N BEING A POSITIVE INTEGER

We have $2 \cos \theta = x + 1/x$

$\Rightarrow (2 \cos \theta)^n = (x + 1/x)^n$

$$\Rightarrow\ 2^n \cos^n \theta = x^n + {}^nc_1\, x^{n-1}\, \frac{1}{x} + {}^nc_2\, x^{n-2}\, \frac{1}{x^2} + {}^nc_3\, x^{n-3}\, \frac{1}{x^3} + \ldots$$

$$\ldots + {}^nc_{n-2}.x^2\, \frac{1}{x^{n-2}} + {}^nc_{n-1}\, x.\frac{1}{x^{n-1}} + {}^nc_n\, \frac{1}{x^n}$$

But ${}^nc_r = {}^nc_{n-r}$, ${}^nc_1 = {}^nc_{n-1}$ etc. and ${}^nc_n = 1$

$\therefore 2^n \cos^n \theta = x^n + {}^nc_1\, x^{n-2} + {}^nc_2\, x^{n-4} + n^{n-4} + {}^nc_3\, x^{n-6} + \ldots.$

$$\ldots + {}^nc_2\, \frac{1}{x^{n-4}} + {}^nc_1\, \frac{1}{x^{n-2}} + \frac{1}{x^n} \qquad \ldots(i)$$

$$= \left(x^n + \frac{1}{x^n}\right) + {}^nc_1 \left(x^{n-2} + \frac{1}{x^{n-2}}\right) + {}^nc_2 \left(x^{n-4} + \frac{1}{x^{n-4}}\right) + \ldots,$$

arranging the terms taking 1st and last, 2nd and last but one and soon.

$$\Rightarrow 2^n \cos^n \theta = (2 \cos n\theta) + {}^nc_1\, [2 \cos (n-2)\, \theta]$$

$$+ {}^nc_2\, [2 \cos (n-4)\, \theta] + \ldots, \qquad \ldots(ii)$$

substituting the values of $x^n + \dfrac{1}{x^n}$ etc.

$$\Rightarrow 2^{n-1} \cos^n \theta = \cos n\theta + {}^nc_1 \cos (n-2)\, \theta + {}^nc_2 \cos (n-4)\, \theta + \ldots \qquad \ldots(iii)$$

If n be odd, the number of terms in R.H.S. of (i), will be (n + 1) which is even, so that all the terms form pairs, and the two middle terms being $\dfrac{n+1}{2}$th and $\dfrac{n+3}{2}$th will form the last pair

$$\frac{{}^nc_{n-1}}{2}\left(x + \frac{1}{x}\right), \text{ since } \frac{{}^nc_{n+1}}{2} = \frac{{}^nc_{n-1}}{2},$$

Therefore the last term in (ii) $= \dfrac{{}^nc_{n-1}}{2}\, (2 \cos \theta)$

$$\Rightarrow \text{the last term in (iii)} = \frac{{}^nc_{n-1}}{2} \cos \theta = \frac{n!\cos q}{\{(n-1)/2\,!\}\,\{(n+1)/2\,!\}}$$

Again if n be even, the number of terms in R.H.S. of (i) will be (n + 1) which is odd, so that all the terms form pair except the middle one in (i), which is independent of x and is equal to ${}^nc_{n/2}$ as it is the $\left(\frac{1}{2}n + 1\right)$ th term in the expansion on R.H.S. of (i). Hence the last term in (iii)

$$= {}^nc_{n/2} = \frac{n}{(n/2)\,!\,(n/2)\,!} = \frac{n\,!}{\{(n/2)\,!\}^2}$$

Example 1:

Expand cos^6 θ.

Solution:

If $x = \cos\theta + i \sin\theta$, then $1/x = \cos\theta - i\sin\theta$;

$x^n = \cos n\theta + i \sin n\theta$ and $1/x^n = \cos n\theta - i \sin n\theta$.

Hence $x + (1/x) = 2\cos\theta$ and $x^n + (1/x^n) = 2\cos n\theta$.

Now $(2\cos\theta)^6 = \left(x+\frac{1}{x}\right)^6$

$$\Rightarrow 2^6 \cos^8\theta = x^6 + 6x^4 + 15x^2 + 20 + 15.\frac{1}{x^2} + 6\frac{1}{x^4} + \frac{1}{x^6}$$

$$\Rightarrow 2^6 \cos^6\theta = \left(x^6+\frac{1}{x^6}\right)+6\left(x^4+\frac{1}{x^4}\right)+15\left(x^2+\frac{1}{x^2}\right)+20$$

$$= (2\cos 6\theta) + 6(2\cos 4\theta) + 15(2\cos 2\theta) + 20,$$

$$\therefore x^n + \frac{1}{x^n} = 2\cos n\theta$$

$2^5 \cos^6\theta = \cos 6\theta + 6\cos 4\theta + 15\cos 2\theta + 10$ *(Kashmir 74)*

$$\Rightarrow \cos^6\theta = \frac{1}{22}\cos 6\theta + \frac{33}{16}\cos 4\theta + \frac{15}{22}\cos 2\theta + \frac{5}{16}$$ **Ans.**

Example 2:

Expand cos^9 θ.

Solution:

With the same o'ratio as in above, we have $(2\cos\theta)^9 = (x + 1/x)^9$

$$= x^9 + 9c_1 x^8 (1/x) + 9c_2 x^7 (1/x)^2 + 9c_3 x^6 (1/x)^3$$
$$+ 9c_4x^5 (1/x)^4 + 9c_5x^4 (1/x)^5 + 9c_0x^3 (1/x)^6$$
$$+ 9c_7x^2 (1/x)^7 + 9c_8x (1/x)^8 + 9c_9 (1/x)^9$$

$$= x^9 + 9x^7 + 36x^5 + 84x^3 + 126x + \frac{126}{x}+\frac{84}{x^5}+\frac{36}{x^5}+\frac{9}{x^7}+\frac{1}{x^9}$$

$$= \left(x^9+\frac{1}{x^9}\right)+9\left(x^7+\frac{1}{x^7}\right)+36\left(x^5+\frac{1}{x^5}\right)$$
$$+ 84\left(x^3+\frac{1}{x^3}\right)+126\left(x+\frac{1}{x}\right)$$

$= (2 \cos 9\theta) + 9(2 \cos 7\theta) + 36(2\cos 5\theta) + 84 \cos (2 \cos 3\theta) + 126 (2 \cos\theta)$

$\cos^9\theta = (1/2)^8 [\cos 9\theta + 9 \cos 7\theta + 36 + 50 + 84 \cos 3\theta + 126 \cos \theta]$

Ans.

Example 3:

Expand $\cos^8 \theta$.

Solution:

With the same notations as in we get $(2 \cos \theta)^8 = \left(x + \frac{1}{x}\right)^8$

$= x^8 + {}^8c_1\, x^7 (1/x) + {}^8c_2 (1/x^2) + {}^8c_3 x^5 (1/x^3)$
$+ {}^8c_4\, x^4 (1/x^4) + {}^8c_5\, x^3 (1/x^5) + {}^8c_6 x^2 (1/x^6)$
$+ {}^3c_7\, x (1/x^7) (1/x^7) + {}^8c_8 (1/x)^8$

$= x^8 + 8x^6 + 28x^4 + 56x^2 + 70 + 56 (1/x^2)$
$+ 28 (1/x^4) + 8 (1/x^7) + {}^8c_8 (1/x)^8$

$$= \left(x^8 + \frac{1}{x^8}\right) + 8\left(x^6 + \frac{1}{x^6}\right) + 28\left(x^4 + \frac{1}{x^4}\right) + 56\left(x^2 + \frac{1}{x^2}\right) + 70$$

$\Rightarrow 2^8 \cos^8 \theta = (2 \cos 8\theta) + 8 (2 \cos 6\theta) + 28 (2 \cos 4\theta)$
$+ 56 (2 \cos 2\theta) + 70$

$\Rightarrow \cos^8 = \frac{1}{2^7} [\cos 8\theta + 8 \cos 6\theta + 28 \cos 4\theta + 56 \cos 2\theta + 35]$ **Ans.**

5.3 EXPANSION OF SIN^N Θ IN A SERIES OF COSINES AND SINES OF MULTIPLES OF Θ, ACCORDING AS n (A POSITIVE INTEGER) IS EVEN OR ODD.

We have $(2i \sin \theta)^n = (x - 1/x)^n$

$\Rightarrow 2^n i^n \sin^n \theta = x^n - {}^nc_1\, x^{n-1} (1/x) + {}^nc_2\, x^{n-2} (1/x^2) - \dots$
$\dots + (-1)^n\, {}^nc_n (1/x^n)$

$\Rightarrow 2^n i^n \sin^n \theta = x^n - {}^nc_1\, x^{n-2} + {}^nc_2\, x^{n-4} - \dots$
$\dots + (-1)^n\, {}^nc_n (1/x^n) (1/x^n)$ ...(i)

Case I n is even: In this case the number of terms in R.H.S. of (i) will be odd and hence the last term is positive, last but one is negative and so on.

Also the middle term which is independent of x is $(-1)^{n/2}\, {}^nc_{n/2}$

Also $i^n = (-1)^{n/2}$

Now as ${}^nc_r = {}^nc_{n-r}$, ${}^nc_{n-1} = {}^nc_1$, etc and ${}^nc_n = 1$,

∴ from (i) after taking the terms in pairs as in we have

$$2^n (-1)^{n/2} \sin^n \theta = \left(x^n + \frac{1}{x^n}\right) - {}^nc_1 \left(x^{n-2} + \frac{1}{x^{n-2}}\right)$$

$$+ {}^nc_2 \left(x^{n-4} + \frac{1}{x^{n-4}}\right) - ... + (-1)^{n/2}\, {}^nc_{n/2}$$

$$\Rightarrow 2^n (-1)^{n./2} \sin^n \theta = (2 \cos n\theta) - {}^nc_1 [2 \cos (n-2)\theta]$$

$$+ {}^nc_2 [2 \cos (n-4)\theta] - ... + (-1)^{n/2}\, {}^nc_{n/2}$$

$$\Rightarrow 2^{n-1} (-1)^{n/2} \sin^n \theta = \cos n\theta - {}^nc_1 \cos (n-2)\theta$$

$$+ {}^nc_2 \cos (n-4)\theta - ... + \frac{(-1)^{n/2}}{2}\, {}^nc_{n/2}$$

Case II – n is odd: In this case the number of terms in R.H.S. of (i) will be even and hence the last term is negative, last but one is positive and so on. Also the two middle terms are

$$(-1)^{(n-1)/2}\, {}^nc_{(n-1)/2}\, x \text{ and } (-1)^{(n+1)/2}\, {}^nc_{(n+1)/2}\, \frac{1}{x}.$$

But ${}^nc_{(n-1)/2} = {}^nc_{(n+1)/2}$, hence these two middle terms taken in a pair = $(-1)^{(n-1)/2}\, {}^nc_{(n-1)/2}\, (x - 1/x)$.

Hence, in this case, from (i) after combining terms in pairs as in we have

$$2^n i^n \sin^n \theta = \left(x^n - \frac{1}{x^n}\right) - {}^nc_1 \left(x^{n-2} - \frac{1}{x^{n-2}}\right)$$

$$+ {}^nc_2 \left(x^{n-4} - \frac{1}{x^{n-4}}\right) + ... + (-1)^{(n-1)/2}\, {}^nc_{(n-1)/2} \times (x - 1/x)$$

$$= (2i \sin n\theta) - {}^nc_2 [2i \sin (n-2)\theta]$$

$$+ {}^nc_4 [2i \text{ s n } (n-4)\theta] + ... + (-1)^{(n-1)/2}\, {}^nc_{(n-1)/2} \times (2i \sin \theta)$$

$$\Rightarrow 2^{(n-1)/2} i^{n-1} \sin^n \theta = \sin n\theta - {}^nc_1 \sin (n-2)\theta + {}^nc_2 \sin (n-4)\theta -$$

$$... + \frac{(-1)^{(n-1)/2}}{2} .{}^nc_{(n-1)/2} \sin \theta$$

$$\Rightarrow 2^{n-1} (-1)^{n-1/2} \sin^n \theta = \sin n\theta - {}^nc_1 \sin (n-2)\theta$$

$$+ {}^nc_2 \sin (n-4)\theta - ... + \frac{(-1)^{(n-1)/2}}{2}\, {}^nc_{(n-1)/2} \sin \theta$$

Example 1:

Expand $\sin^7 \theta$.

Solution:

Let $x = \cos \theta + i \sin \theta$, then $(1/x) = \cos \theta - i \sin \theta$,

$x^n = \cos n\theta + i \sin n\theta$, $(1/x^n) = \cos n\theta - i \sin n\theta$,

whence $x - 1/x = 2i \sin \theta$ and $x^n - 1/x^n = 2i \sin n\theta$,

Now $(2i \sin \theta)^7 = \left(x - \frac{1}{x}\right)^7$

$$= x^7 - {}^7c_1\, x^6.\frac{1}{x} + {}^7c_2 x^5.\frac{1}{x^2} - {}^7c_3 x^4.\frac{1}{x^3} + {}^7c_4 x^3.\frac{1}{x^4}$$

$$- {}^7c_5 x^2.\frac{1}{x^5} + {}^7c_6 x.\frac{1}{x_6} - {}^7c_7 \frac{1}{x^7}$$

$$= x^7 - 7x^5 + 21x^3 - 35x + \frac{35}{x} - \frac{21}{x^3} + \frac{7}{x^5} - \frac{1}{x^7}$$

$$\Rightarrow 2^7 i^7 \sin^7 \theta = \left(x^7 - \frac{1}{x^7}\right) - 7\left(x^5 - \frac{1}{x^5}\right) + 21\left(x^3 - \frac{1}{x^3}\right) - 35\left(x - \frac{1}{x}\right)$$

$$\Rightarrow -2^7 i \sin^7 \theta = (2i \sin 7\theta) - 7\,(2i \sin 5\theta) + 21\,(2i \sin 3\theta) - 35\,(2i \sin\theta)$$

$$\Rightarrow -\,2^6 \sin^7 \theta = \sin 7\theta - 7 \sin 5\theta + 21 \sin 3\theta - 35 \sin \theta$$

$$\Rightarrow \sin^7 \theta = -\,\frac{1}{2^6}\,[\sin 7\theta - 7 \sin 5\theta + 21 \sin 3\theta - 35 \sin \theta]$$ **Ans.**

Example 2:

Expand $\sin^8 \theta$.

Solution:

With the same notations as in above we have

$x + 1/x = 2 \cos \theta$; $x - 1/x = 2i \sin \theta$; $x^n + 1/x^n = 2 \cos n\theta$

Now $(2i \sin \theta)^8 = (x - 1/x)^8$

$$\Rightarrow 2^8 i^8 \sin^8 \theta = x^8 - {}^8c_1\, x^7\, \frac{1}{x} + {}^8c_2\, x^5\, \frac{1}{x^3}$$

$$+ {}^8c_4 x^4.\frac{1}{x^4} - {}^8c_5 x^3\, x^3\, \frac{1}{x^5} + {}^8c_6 x^2.\frac{1}{x^6} - {}^8c_7\, x\, \frac{1}{x^7} + {}^8c_8\, \frac{1}{x^8}$$

$$= c^8 - 8x^6 + 28x^4 - 56x^2 + 70 - \frac{56}{x^2} + \frac{28}{x^4} - \frac{8}{x^6} + \frac{1}{x^8}$$

$$\Rightarrow 2^8 \sin^8 \theta = \left(x^8 + \frac{1}{x^8}\right) - 8\left(x^6 + \frac{1}{x^6}\right) + 28\left(x^4 + \frac{1}{x^4}\right) 56\left(x^2 + \frac{1}{x^2}\right) + 70$$

$$= (2 \cos 8\theta) - 8\,(2 \cos 6\theta) + 28\,(2 \cos 4\theta) - 56\,(2 \cos 2\theta) + 70$$

$$\Rightarrow \sin^8 \theta = \frac{1}{2^7}\,[\cos 8\theta - 8 \cos 6\theta + 28 \cos 4\theta - 56 \cos 2\theta + 35]$$ **Ans.**

Example 3:

Expand $\sin^9 \theta$.

Solution:

With the same notations as in above we have

$$(2i \sin \theta)^9 = (x - 1/x)^9$$

$$\Rightarrow 2^9 i^9 \sin^9 \theta = x^9 - {}^9c_1 x^8.\frac{1}{x} + {}^9c_2\, x^7.\frac{1}{x^2} - {}^9c_3\, x^6.\frac{1}{x^3}$$

$$+ {}^9c_4 x^5.\frac{1}{x^4} - {}^9c_5 x^4.\frac{1}{x^5} + {}^9c_6 x^3.\frac{1}{x^6} - {}^9c_7\, x^2.\frac{1}{x^7}$$

$$+ {}^9c_8\, x.\frac{1}{x^8} - {}^9c_9\,\frac{1}{x^9}.$$

$$\Rightarrow 2^9 i \sin^9 \theta = x^9 - 9x^7 + 36x^5 - 84x^3 + 126x - \frac{126}{x} + \frac{84}{x^3} - \frac{36}{x^5} + \frac{9}{x^7} - \frac{1}{x^9}.$$

$$= \left(x^9 - \frac{1}{x^9}\right) - 9\left(x^7 - \frac{1}{x^7}\right) + 36\left(x^5 - \frac{1}{x^5}\right)$$

$$-84\left(x^3 - \frac{1}{x^3}\right) - 126\left(x - \frac{1}{x}\right)$$

$$\Rightarrow \quad 2^9 i \sin^9 \theta = (2i \sin 9\theta) - 9\,(2i \sin 7\theta) + 36\,(2i \sin 5\theta)$$

$$- 84\,(2i \sin 3\theta) + 126\,(2i \sin \theta).$$

$$\Rightarrow \sin^9 \theta = \frac{1}{2^9}\,[\sin 9\theta - 9 \sin 7\theta + 36 \sin 5\theta - 84 \sin 3\theta + 126 \sin\theta].$$

Ans.

Example 4:

Express $\cos^5 \theta \sin^3 \theta$ in terms of since of multiple of θ.

Solution:

Let $x = \cos \theta + i \sin \theta$, then $1/x = \cos \theta - i \sin \theta$,

$x^n = \cos n\theta + i \sin n\theta$ and $1/x^n = \cos n\theta - i \sin n\theta$,

whence $x + 1/x = 2 \cos \theta$, $x - 1/x = 2i \sin \theta$ etc.

Now $(2 \cos \theta)^5 (2i \sin \theta)^3 = (x + 1/x)^5 (x - 1/x)^3$ ...(i)

If we actually expand and multiply R.H.S. of (i), then the process becomes very lengthy and laborious. In order to avoid the same the following method is used to find out the coefficients of the products. **(Note)**

Coefficients of $\left(x + \frac{1}{x}\right)^5$

1 5 10 10 5 1

Multiplying by $\left(x-\frac{1}{x}\right)$, once, the coefficients are

1 (5 – 1) (10 – 5) (10 – 10) (5 – 10) (1 – 5) (0 – 1) **(Note)**

i.e., 1 4 5 0 – 5 – 4 – 1

Multiplying twice, coefficients are

1 3 1 –5 – 5 1 3 –1

Multiplying thrice, coefficients are

1 2 –2 – 6 0 6 2 – 1 – 1

Hence from (i) we have

$$(2\cos\theta)^5 (2i\sin\theta)^3 = x^8 + 2x^6 - 2x^4 - 6x^2 + 0 + \frac{6}{x^2} + \frac{2}{x^4} - \frac{2}{x^6} - \frac{1}{x^8}$$

$$\Rightarrow 2^8 i^3 \cos^5\theta \sin^3\theta = \left(x^8 - \frac{1}{x^8}\right) + 2\left(x^6 - \frac{1}{x^6}\right) - 2\left(x^4 - \frac{1}{x^4}\right) - 6\left(x^2 - \frac{1}{x^2}\right)$$

$$\Rightarrow -2^8 i \cos^5\theta \sin^3\theta = (2i\sin 8\theta) + 2(2i\sin 6\theta) - 2(2i\sin 4\theta) - 6(2i\sin 2\theta)$$

$$\Rightarrow -2^7 \cos^5\theta \sin^3\theta = \sin 8\theta + 2\sin 6\theta - 2\sin 4\theta - 6\sin 2\theta$$

$$\Rightarrow \cos^5\theta \sin^3\theta = -\frac{1}{128}[\sin 8\theta + 2\sin 6\theta - 2\sin 4\theta - 6\sin 2\theta].$$

Ans.

Example 5:

Prove that

$$32\sin^4\theta\cos^2\theta = \cos 6\theta - 2\cos 4\theta - \cos 2\theta + 2$$

Solution:

With the same notations as in above, we have

$$(2i\sin\theta)^4 (2\cos\theta)^2 = \left(x-\frac{1}{x}\right)^4 \left(x+\frac{1}{x}\right)^2 \quad ...(i)$$

In order to find the coefficients of the product in R.H.S. of (i) we have as follows:

Coefficients of $\left(x-\frac{1}{x}\right)$ with proper sign are

1 – 4 6 – 4 1

Multiplying by $\left(x+\frac{1}{x}\right)$ once, the coefficients are

1 (– 4 + 1) (6 – 4) (– 4 + 6) (1 – 4) (0 + 1) **(Note)**

i.e., 1 – 3 2 2 – 3 1

Multiplying by $\left(x+\frac{1}{x}\right)$ twice, coefficients are

1 – 2 – 1 4 – 1 – 2 1

Hence from (i) we have

$$2^4 i^4 \sin^4 \theta\, 2^2 \cos^2 \theta = x^6 - 2x^4 - x^2 + 4 - \frac{1}{x^2} - \frac{2}{x^4} + \frac{1}{x^6}$$

$$\Rightarrow 2^6 \sin^4 \theta \cos^2 \theta = \left(x^6+\frac{1}{x^6}\right) - 2\left(x^4+\frac{1}{x^4}\right) - \left(x^2+\frac{1}{x^2}\right) + 4$$

$$\Rightarrow 64 \sin^4 \theta \cos^2 \theta = (2 \cos 6\theta) - 2\,(2 \cos 4\theta) - (2 \cos 2\theta) + 4$$

$\Rightarrow 32 \sin^4 \theta \cos^2 \theta = \cos 6\theta - 2 \cos 4\theta - \cos 2\theta + 2$. **Hence proved.**

Example 6:

Prove that

$$-256 \sin^7 \theta \cos^2 \theta = \sin 9\theta - 5 \sin 7\theta + 8 \sin 5\theta - 14 \sin \theta.$$

Solution:

With the same notations as in we have

$$(2i \sin \theta)^7 (2 \cos \theta)^2 = \left(x-\frac{1}{x}\right)^7 \left(x+\frac{1}{x}\right)^2 \quad \text{...(i)}$$

In order to find the coefficients of the product in R.H.S. of (i) we proceed as follows:

Coefficients of $\left(x-\frac{1}{x}\right)$ with proper sign are

1 – 7 21 – 35 35 – 21 7 –1

Multiplying by $\left(x+\frac{1}{x}\right)$ the coefficients are

1 – 6 14 – 14 0 14 – 14 6 –1

Multiplying by $\left(x+\frac{1}{x}\right)$ twice, the coefficients are

1 –5 8 0 – 14 14 0 – 8 5 –1

Hence from (i) we have

$$2^7\ i^7 \sin^7\theta\ 2^2\cos^2\theta = x^9 - 5x^7 + 8x^5 + 0x^3 - 14x + \frac{14}{x} + \frac{0}{x^3} - \frac{8}{x^5} + \frac{5}{x^7} - \frac{1}{x^9}$$

$$\Rightarrow -2^9\ i \sin^7\theta \cos^2\theta = \left(x^9 - \frac{1}{x^9}\right) - 5\left(x^7 - \frac{1}{x^7}\right) + 8\left(x^5 - \frac{1}{x^5}\right) - 14\left(x - \frac{1}{x}\right)$$

$$= (2i \sin 9\theta) - 5\,(2i \sin 7\theta) + 8\,(2i \sin 5\theta) - 14\,(2i \sin \theta)$$

$$\Rightarrow -2^8 \sin^7\theta \cos^2\theta = \sin 9\theta - 5 \sin 7\theta + 8 \sin 5\theta - 14 \sin\theta$$

$$\Rightarrow -256 \sin^7\theta \cos^2\theta = \sin 9\theta - 5 \sin 7\theta + 8 \sin 5\theta - 14 \sin\theta.$$

Hence proved.

Example 7:

Express $\sin^6\theta \cos^2\theta$ in terms of cosines of multiples of θ.

Solution:

With the same notations as in we have

$$(2i \sin\theta)^6\ (2\cos\theta)^2 = \left(x - \frac{1}{x}\right)^6 \left(x + \frac{1}{x}\right)^2 \qquad ..(i)$$

In order to find the coefficients of the product in R.H.S. of (i) we proceed as follows:

Coefficients of $\left(x - \frac{1}{x}\right)^6$ with proper sign are

1 – 6 15 – 20 15 – 6 1

Multiplying by $\left(x + \frac{1}{x}\right)$ the coefficients are

1 – 5 9 – 5 – 5 9 – 5 1

Multiplying by $\left(x + \frac{1}{x}\right)$ twice, the coefficients are

1 – 4 4 4 – 10 4 4 – 4 1

Hence from (i) we have

$$2^6\ i^6 \sin^6\theta\ 2^2\cos^2\theta = x^8 - 4x^6 + 4x^4 + 4x^2 - 10 + \frac{4}{x^2} + \frac{4}{x^4} - \frac{4}{x^6} + \frac{1}{x^8}$$

$$\Rightarrow -2^8 \sin^6\theta \cos^2\theta = \left(x^8 + \frac{1}{x^8}\right) - 4\left(x^6 + \frac{1}{x^6}\right)$$
$$+ 4\left(x^4 + \frac{1}{x^4}\right) + 4\left(x^2 + \frac{1}{x^2}\right) - 10$$

$$\Rightarrow -2^8 \sin^6\theta \cos^2\theta = (2\cos 8\theta) - 4(2\cos 6\theta)$$
$$+ 4(2\cos 4\theta) + 4(2\cos 2\theta) - 10$$

$$\Rightarrow -2^7 \sin^6\theta \cos^2\theta = \cos 8\theta - 4\cos 6\theta + 4\cos 4\theta + 4\cos 2\theta - 5$$

$$\Rightarrow \sin^6\theta \cos^2\theta = -\frac{1}{2^7}[\cos 8\theta - 4\cos 6\theta + 4\cos 4\theta + 4\cos 2\theta - 5].$$

Ans.

Example 8:

Express $\cos^5\theta \sin^7\theta$ *in a series of sines of multiple of* θ.

Solution:

With the previous notations, we have

$$(2i\sin\theta)^7 (2\cos\theta)^5 = \left(x - \frac{1}{x}\right)\left(x + \frac{1}{x}\right)^5 \quad \text{...(i)}$$

In order to find the coefficients of the product in R.H.S. of (i), we proceed as follows

Coefficients of $\left(x - \frac{1}{x}\right)$ with proper sign are

1 – 7 21 – 35 35 – 21 7 – 1

Multiplying by $\left(x + \frac{1}{x}\right)$, the coefficients are

1 – 6 14 –14 0 14 – 14 6 –1

Multiplying by $\left(x + \frac{1}{x}\right)$ twice, coefficients are

1 – 5 8 0 – 14 0 – 8 5 –1

Multiplying by $\left(x + \frac{1}{x}\right)$ thrice, coefficients are

1 – 4 3 8 – 14 0 14 – 8 – 3 4 – 1

Multiplying by $\left(x + \frac{1}{x}\right)$ four times, coefficients are

1 – 3 –1 11 – 6 – 14 6 11 1 3 – 1

Multiplying by $\left(x+\frac{1}{x}\right)$ five times, coefficients are

1 −2 −4 10 5 −20 0 20 −5 −10 4 2 −1

Hence from (i) we have

$$2^7 i^7 \sin^7 \theta\, 2^5 \cos^5 \theta = x^{12} - 4x^8 + 10x^6 + 5x^4 - 20x^2 + 0$$
$$+\frac{20}{x^2} - \frac{5}{x^4} - \frac{10}{x^6} + \frac{4}{x^8} + \frac{2}{x^{10}} - \frac{1}{x^{12}}$$

$$\Rightarrow -2^{12} i \sin^7 \theta \cos^5 \theta = \left(x^{12} - \frac{1}{x^{12}}\right) - \left(x^{10} - \frac{1}{x^{10}}\right)$$
$$-4\left(x^8 - \frac{1}{x^8}\right) + 10\left(x^6 - \frac{1}{x^6}\right) + 5\left(x^4 - \frac{1}{x^4}\right) - 20\left(x^2 - \frac{1}{x^2}\right)$$
$$= (2i \sin 12\theta) - 2\,(2i \sin 10\theta) - 4\,(2i \sin 8\theta)$$
$$+ 10\,(2i \sin 6\theta) + 5\,(2i \sin 4\theta) - 20\,(2i \sin 2\theta)$$

$$\Rightarrow -2^{11} \sin^7 \theta \cos^5 \theta = \sin 12\theta - 2 \sin 10\theta - 4 \sin 8\theta$$
$$+ 10 \sin 6\theta + 5 \sin 4\theta - 20 \sin 2\theta$$

$$\Rightarrow \sin^7 \theta \cos^5 \theta = -\frac{1}{2^{11}}\,[\sin 12\theta - 2 \sin 10\theta - 4 \sin 8\theta$$
$$+ 10 \sin 6\theta + 5 \sin 4\theta - 20 \sin 2\theta]$$ **Ans.**

5.4 EXPANSIONS OF COS^N Θ AND SIN^N Θ IN TERMS OF SINES AND COSINES OF MULTIPLES OF Θ, N BEING A POSITIVE INTEGER

Let $z = \cos \theta + i \sin \theta$ so that $\frac{1}{z} = \cos \theta - i \sin \theta$.

Now $z + \frac{1}{z} = 2 \cos \theta$ and $z - \frac{1}{z} = 2i \sin \theta$.

By Expansion of Some Trigonometrical Functions, we have

$z^n = \cos n\theta + i \sin n\theta$ and $\frac{1}{z^n} = \cos n\theta - i \sin n\theta$.

Thus $z^n + \frac{1}{z^n} = 2 \cos n\theta$ and $z^n - \frac{1}{z^n} = 2i \sin n\theta$. ...(2)

From (2), we have

$$(2 \cos \theta)^n = \left(z + \frac{1}{z}\right)^n.$$

Expanding the R.H.S. with the help of the Binomial Theorem and using (2), we get the required expansion of $\cos^n \theta$. Similarly

$$(2i \sin \theta)^n = \left(z - \frac{1}{z}\right)^n$$

gives the required expansion of $\sin^n \theta$.

Example 1:

Prove that

(i) $16 \sin^5 \theta = \sin 5\theta - 5 \sin 3\theta + 10 \sin \theta$.

(ii) $16 \cos^5 \theta = \cos 5\theta + 5 \cos 3\theta + 10 \cos \theta$.

Solution:

Let $z = \cos \theta + i \sin \theta$ so that $1/z = \cos \theta - i \sin \theta$.

$$\Rightarrow z - \frac{1}{z} = 2i \sin \theta$$

$$\Rightarrow (2i \sin \theta)^5 = \left(z - \frac{1}{z}\right)^5$$

$$= z^5 - {}^5C_1\, z^4 . \frac{1}{z} + {}^5C_2\, z^3 . \frac{1}{z^2} - {}^5C_3 z^2 . \frac{1}{z^3} + {}^5C_4 z . \frac{1}{z^4} - \frac{1}{z^5}$$

$$= z^5 - 5\, z^3 + \frac{5.4}{2.1} z - \frac{5.4.3}{3.2.1} . \frac{1}{z} + \frac{5.4.3.2}{4.3.2.1} . \frac{1}{z^3} - \frac{1}{z^5}$$

$$= \left(z^5 - \frac{1}{z^5}\right) - 5\left(z^3 - \frac{1}{z^3}\right) + 10\left(z - \frac{1}{z}\right).$$

$= (2i \sin 5\theta) - 5\,(2i \sin 3\theta) + 10\,(2i \sin \theta)$, by (2) above

$\Rightarrow 32i \sin^5 \theta = 2i\,(\sin 5\theta - 5 \sin 3\theta + 10 \sin \theta)$ $\quad (\because i^5 = i)$

Hence $16 \sin^5 \theta = \sin 5\theta - 5 \sin 3\theta + 10 \sin \theta$.

(ii) We have $z + 1/z = 2 \cos \theta$, and so

$$(2 \cos \theta)^5 = \left(z + \frac{1}{z}\right)^5 = z^5 + 5z^3 + 10z + 10\frac{1}{z} + 5\frac{1}{z^3} + \frac{1}{z^5},$$

(as shown in the first part)

$$= \left(z^5 + \frac{1}{z^5}\right) + 5\left(z^3 + \frac{1}{z^3}\right) + 10\left(z + \frac{1}{z}\right)$$

$\Rightarrow 2^5 \cos^5 \theta = (2 \cos 5\theta) + 5\,(2 \cos 3\theta) + 10\,(2 \cos \theta)$, by (2) above.

Hence $16 \cos^5 \theta = \cos 5\theta + 5 \cos 3\theta + 10 \cos \theta$.

Example 2:

Prove that $32 \cos^6 \theta = \cos 6\theta + 6 \cos 4\theta + 15 \cos 2\theta + 10$.

Solution:

Let $x = \cos\theta + i\sin\theta$ so that $1/x = \cos\theta - i\sin\theta$.

Thus $2\cos\theta = x + \dfrac{1}{x} \Rightarrow (2\cos\theta)^6 = \left(x+\dfrac{1}{x}\right)^6$

$$= x^6 + {}^6C_1x^5 \cdot \frac{1}{x} + {}^6C_2x^4 \cdot \frac{1}{x^2} + {}^6C_3x^3 \cdot \frac{1}{x^3} + {}^6C_4x^2 \cdot \frac{1}{x^4} + {}^6C_5x \cdot \frac{1}{x^5} + \frac{1}{x^6}$$

$$= x^6 + 6x^4 + 15x^2 + + 20 + \frac{15}{x^2} + \frac{6}{x} + \frac{1}{x^6}$$

$$= \left(x^6 + \frac{1}{x^6}\right) + 6\left(x^4 + \frac{1}{x^4}\right) + 15\left(x^2 + \frac{1}{x^2}\right) + 20$$

$\therefore\ 64\cos^6\theta = 2\cos 6\theta + 6.2\cos 4\theta + 15.2\cos 2\theta + 20$

Hence $32\cos^6\theta = \cos 6\theta + 6\cos 4\theta + 15\cos 2\theta + 10$.

Example 3:

Prove that

$$64\cos^7\theta = \cos 7\theta + 7\cos 5\theta + 21\cos 3\theta + 35\cos\theta.$$

Solution:

Let $x = \cos\theta + i\sin\theta$, then $1/x = \cos\theta - i\sin\theta$.

$$\therefore (2\cos\theta)^7 = \left(x+\frac{1}{x}\right)^7 = x^7 + 7c_1.x^6 \cdot \frac{1}{x} + 7\,c_2.x^5 \cdot \frac{1}{x^2}$$

$$+ 7c_3x^4 \cdot \frac{1}{x^3} + 7c_4.x^3 \cdot \frac{1}{x^4} + 7c_5x^2. + 7c_6x. + 7c_7 \cdot \frac{1}{x^7}$$

$$= x^7 + 7x^5 + 21x^3 + 35x + 35 \cdot \frac{1}{x} + 21 \cdot \frac{1}{x^3} + 7 \cdot \frac{1}{x^5} + \frac{1}{x^7}$$

$$= \left(x^7 + \frac{1}{x^7}\right) + 7\left(x^5 + \frac{1}{x^5}\right) + 21\left(x^3 + \frac{1}{x^3}\right) + 35\left(x + \frac{1}{x}\right)$$

$= 2\,[\cos 7\theta + 7\cos 5\theta + 21\cos 3\theta + 35\cos\theta]$.

Hence $64\cos^7\theta = \cos 7\theta + 7\cos 5\theta + 21\cos 3\theta + 35\cos\theta$.

Example 4:

Prove that

$$\sin^8\theta = \frac{1}{128}\{\cos 8\theta - 8\cos 6\theta + 28\cos 4\theta - 56\cos 2\theta + 70\}.$$

Solution:

Let $z = \cos\theta + i\sin\theta \Rightarrow z^{-1} = \cos\theta - i\sin\theta$

$\Rightarrow z - z^{-1} = 2i\sin\theta \Rightarrow \sin^8\theta = \dfrac{1}{(2i)^8}(z - z^{-1})^8$

$$= \frac{1}{256}\left(z^8 - {}^8c_1 z^6 + {}^8c_2 z^4 - {}^8c_3 z^2 + {}^8c_4 - {}^8c_5 z^{-2} + {}^8c_6 z^{-4} - {}^8c_7 z^{-6} + z^{-8}\right)$$

$$= \frac{1}{256}\{(z^8 + z^{-8}) - 8(z^6 + z^{-6}) + 28(z^4 + z^{-4}) - 56(z^2 + z^{-2}) + 70\}$$

$$= \frac{1}{128}\{\cos 8\theta - 8\cos 6\theta + 28\cos 4\theta - 56\cos 2\theta + 70\}.$$

Example 5:

Show that

$$\sin^6\theta = -\frac{1}{32}[\cos 6\theta - 6\cos 4\theta + 15\cos 2\theta - 10].$$

Solution:

We have $(2i\sin\theta)^6 = \left(x - \dfrac{1}{x}\right)^6$

$$= x^6 + 6c_1x^5\left(-\frac{1}{x}\right) + 6c_2x^4\left(-\frac{1}{x}\right)^2 + 6c_3x^3\left(-\frac{1}{x}\right)^3 + 6c_4x^2\left(-\frac{1}{x}\right)^4 + 6c_5x^2\left(-\frac{1}{x}\right)^5 + 6c_6\left(-\frac{1}{x}\right)^6$$

$$= x^6 - 6x^4 + 15x^2 - 20 + 15\frac{1}{x^2} - 6\frac{1}{x^4} + \frac{1}{x^6}$$

$$= \left(x^6 + \frac{1}{x^6}\right) - 6\left(x^4 + \frac{1}{x^4}\right) + 15\left(x^2 + \frac{1}{x^2}\right) - 20$$

$$= [2\cos 6\theta] - 6[2\cos 4\theta] + 15[2\cos 2\theta] - 20$$

Hence $\sin^6\theta = -\dfrac{1}{32}[\cos 6\theta - 6\cos 4\theta + 15\cos 2\theta - 10]$

Example 6:

Prove that: $32\sin^4\theta\cos^2\theta = \cos 6\theta - 2\cos 4\theta - \cos 2\theta + 2.$

Solution:

Let $x = \cos\theta + i\sin\theta \Rightarrow 1/x = \cos\theta - i\sin\theta.$

$\Rightarrow x + 1/x = 2 \cos \theta$ and $x - 1/x = 2i \sin \theta$,

$$\Rightarrow (2i \sin \theta)^4 (2 \cos \theta)^2 = \left(x+\frac{1}{x}\right)^2 \left(x-\frac{1}{x}\right)^4$$

$$= \left(x+\frac{1}{x}\right)^3 \left(x-\frac{1}{x}\right)^2 \left(x-\frac{1}{x}\right)^2$$

$$= \left(x^2-\frac{1}{x^2}\right)^2 \left(x-\frac{1}{x}\right)^2$$

$$= \left(x^4-2+\frac{1}{x^4}\right)\left(x^2-2+\frac{1}{x^2}\right)$$

$$= x^6 - 2x^4 - x^2 + 4 - \frac{1}{x^2} - \frac{2}{x^4} + \frac{1}{x^6}$$

$$= \left(x^6+\frac{1}{x^6}\right) - 2\left(x^4+\frac{1}{x^4}\right) - \left(x^2+\frac{1}{x^2}\right) + 4$$

$$= 2 \cos 6\theta - 2.2 \cos 4\theta - 2 \cos 2\theta + 4.$$

Hence $32 \sin^4 \theta \cos^2 \theta = \cos 6\theta - 2 \cos 4\theta - \cos 2\theta + 2$.

Example 7:

Prove that $128 \cos^3 \theta \sin^5 \theta = \sin 8\theta - 2 \sin 6\theta - 2 \sin 4\theta + 6 \sin 2\theta$.

Solution:

If $z = \cos \theta + i \sin \theta$, then $z^{-1} = \cos \theta - i \sin \theta$.

$\Rightarrow 2 \cos \theta = z + z^{-1}$, $2i \sin \theta = z - z^{-1}$.

$\Rightarrow (2 \cos \theta)^3 (2i \sin \theta)^5 = (z + z^{-1})^3 (z - z^{-1})^5$

$= (z^2 - z^{-2})^3 (z - z^{-1})^2$

$= (z^6 - 3z^2 + 3z^{-2} - z^{-6}) (z^2 - 2 + z^{-2})$

$= (z^8 - z^{-8}) - 2 (z^6 - z^{-6}) - 2 (z^4 - z^{-4}) + 6 (z^2 - z^{-2})$

$2^8 i \cos^3 \theta \sin^5\theta = 2i \sin 8\theta - 2(2i \sin 6\theta) - 2(2i \sin 4\theta) + 6(2i \sin 2\theta)$

Hence $128 \cos^3 \theta \sin^5 \theta = \sin 8\theta - 2 \sin 6\theta - 2 \sin 4\theta + 6 \sin 2\theta$.

Example 8:

Prove that

$64 \sin^3 \theta \cos^4 \theta = - \sin 7\theta - \sin 5\theta + 3 \sin 3\theta + 3 \sin \theta$.

Solution:

$(2\cos\theta)^4 (2i\sin\theta)^3 = (z + z^{-1})^4 (z - z^{-1})^3$

$= (z^4 + 4z^2 + 6 + 4z^{-2} + z^{-4})(z^3 - 3z + 3z^{-1} - z^{-3})$

$= (z^7 - z^{-7}) + (z^5 - z^{-5}) - 3(z^3 - z^{-3}) - 3(z - z^{-1})$

$\Rightarrow - 2i(64\sin^3\theta\cos^4\theta) = 2i\sin 7\theta + 2i\sin 5\theta - 3(2i\sin 3\theta)$

$- 3(2i\sin\theta)$

Hence $64\sin^2\theta\cos^4\theta = -\sin 7\theta - \sin 5\theta + 3\sin 3\theta + 3\sin\theta$.

Example 9:

Expand $\sin^7\theta\cos^2\theta$ *in a series of sines of multiplies of* θ.

Solution:

Let $z = \cos\theta + i\sin\theta$.

$\therefore (2i\sin\theta)^7 (2\cos\theta)^3 = \left(z - \frac{1}{z}\right)^7 \left(z + \frac{1}{z}\right)^3$

$= \left(z - \frac{1}{z}\right)^4 \left(z^2 - \frac{1}{z^2}\right)^3$

$= (z^4 - 4z^2 + 6 - 4z^{-2} + z^{-4})(z^6 - 3z^2 + 3z^{-2} - z^{-6})$

$= z^{10} - 4z^8 + 3z^6 + 8z^4 - 14z^2 + 14z^{-8} - 8z^{-4} - 3z^{-6} + 4z^{-8} - z^{-10}$

$= (z^{10} - z^{-10}) - 4(z^8 - z^{-8}) + 3(z^6 - z^{-6})$

$+ 8(z^4 - z^{-4}) - 14(z^2 - z^{-2})$

$= 2i[\sin 10\theta - 4\sin 8\theta + 3\sin 6\theta + 8\sin 4\theta - 14\sin 2\theta]$.

Hence $-2^9\sin^7\theta\cos^2\theta = \sin 10\theta - 4\sin 8\theta + 3\sin 6\theta$

$+ 8\sin 4\theta - 14\sin 2\theta$.

Example 10:

Prove that

$-128\sin^6\theta\cos^2\theta = \cos 8\theta - 4\cos 6\theta + 4\cos 4\theta + 4\cos 2\theta - 5$.

Solution:

Let $z = \cos\theta + i\sin\theta$ so that $1/z = \cos\theta - i\sin\theta$.

Thus $z + 1/z = 2\cos\theta$ and $z - 1/z = 2i\sin\theta$.

$\therefore (2i\sin\theta)^6 (2\cos\theta)^2 = \left(z - \frac{1}{z}\right)^6 \left(z - \frac{1}{z}\right)^2$

$$= \left(z - \frac{1}{z}\right)^4 \left(z^2 - \frac{1}{z^2}\right)^2$$

$$= \left(z^4 - 4z^8 . \frac{1}{z} + 6z^2 . \frac{1}{z^2} - 4z . \frac{1}{z^3} + \frac{1}{z^4}\right)\left(z^4 - 2 + \frac{1}{z^4}\right)$$

$$= \left(z^8 + \frac{1}{z^8}\right) - 4\left(z^6 + \frac{1}{z^6}\right) + 4\left(z^4 + \frac{1}{z^4}\right) + 4\left(z^2 + \frac{1}{z^2}\right) + 10$$

$= 2 \cos 8\theta - 4 (2 \cos 6\theta) + 4 (2 \cos 4\theta) + 4 (2 \cos 2\theta) - 10$

$\Rightarrow - 2^8 \sin^6 \cos^2 \theta = 2 (\cos 8\theta - 4 \cos 6\theta + 4 \cos 4\theta + 4 \cos 2\theta - 5)$

Hence $- 128 \sin^6 \theta \cos^2 \theta = \cos 8\theta - 4 \cos 6\theta + 4 \cos 4\theta + 4 \cos 2\theta - 5.$

Example 11:

Prove that

$128 \sin^2 \theta \cos^6 \theta = - 1 \cos 8\theta - 4 \cos 6\theta - 4 \cos 4\theta + 4 \cos 2\theta + 5.$

Solution:

Let $z = \cos \theta + i \sin \theta$ so that $z^{-1} = \cos \theta - i \sin \theta$.

$\therefore z + z^{-1} = 2 \cos \theta, \ z - z^{-1} = 2i \sin \theta.$

We have

$(2i \sin \theta)^2 (2 \cos \theta)^6 = (z - z^{-1})^2 (z + z^{-1})^6$

$= (z - z^{-1})^2 (z + z^{-1})^2 (z + z^{-1})^4$

$= (z^2 - z^{-2})^2 (z^4 + 4c_1 z^3 z^{-1} + 4c_3 z^2 z^{-2} + 4c_3 z z^{-3} + z^{-4})$

$= (z^4 - 2 + z^{-4}) (z^4 + 4z^2 + 6 + 4z^{-2} + z^{-4})$

$= (z^8 + z^{-8}) + 4 (z^6 + z^{-6}) + 4 (z^4 + z^{-4}) - 4 (z^2 + z^{-2}) - 10$

$\therefore - 2^8 \sin^2 \theta \cos^6 \theta = 2 \cos 8\theta + 4 (2 \cos 6\theta)$
$+ 4 (2 \cos 4\theta) - 4 (2 \cos 2\theta) - 10.$

Hence

$128 \sin^2 \theta \cos^6 \theta = - \cos 8\theta - 4 \cos 6\theta - 4 \cos 4\theta + 4 \cos 2\theta + 5.$

Example 12:

Prove that

$256 \sin^5 \theta \cos^4 \theta = \sin 9\theta - \sin 7\theta - 4 \sin 5\theta + 4 \sin 3\theta + 6 \sin \theta.$

Solution:

We have

$(2i \sin \theta)^5 (2 \cos \theta)^4 = (z - z^{-1})^5 (z + z^{-1})^4$

$= (z^2 - z^{-2}) (z - z^{-1})$

$= (z^8 - 4z^4 + 6 - 4z^{-4} + z^{-8}) (z - z^{-1})$

$= (z^9 - z^{-9}) - (z^7 - z^{-7}) - 4 (z^5 - z^{-5}) + 4 (z^3 - z^{-3}) + 6 (z - z^{-1})$

$\therefore\ i\, 2^9 \sin^5 \theta \cos^4 \theta = 2 i \sin 9\theta - 2i \sin 7\theta - 4 (2i \sin 5\theta)$

$+ 4 (2i \sin 3\theta) + (2i \sin \theta).$

Hence

$256 \sin^5 \theta \cos^4 \theta = \sin 9\theta - \sin 7\theta - 4 \sin 5\theta + 4 \sin 3\theta + 6 \sin \theta.$

5.5 EXPRESSION FOR TAN $(\Theta_1 + \Theta_2 + ... + \Theta_N)$ IN TERMS OF TAN Θ_1, TAN Θ_2,..., TAN Θ_N

We have $\cos (\theta_1 + \theta_2 +...+ \theta_n) + i \sin (\theta_1 + \theta_2 + ... + \theta_n)$

$= (\cos \theta_1 + i \sin \theta_1) (\cos \theta_2 + i \sin \theta_2) ... (\cos \theta_n + i \sin \theta_n)$

$= \cos \theta_2 \cos \theta_2 ... \cos \theta_n (1 + i \tan \theta_1) (1 + i \tan \theta_2).....$

$(1 + i \tan \theta_n)$

$= \cos \theta_1 \cos \theta_2 ... \cos \theta_n (1 + iS_1 - S_2 - iS_3 + S_4 + ...),$ ...(1)

where $S_1 = \tan \theta_1 + \tan \theta_2 + ... + \tan \theta_n = S \tan \theta_1,$

$S_2 = \tan \theta_1 \tan \theta_2 + ... = \Sigma \tan \theta_1 \tan \theta_2,$

$S_3 = \Sigma \tan \theta_1 \tan \theta_2 \tan \theta_3$, and so on.

Equating real and imaginary parts in (1), we have

$\cos (\theta_1 + \theta_2 + ... + \theta_n) = \cos \theta_1 \cos \theta_2 ... \cos \theta_n (1 - S_2 + S_4 ...),$

and $\sin (\theta_1 + \theta_2 + ... + \theta_n) = \cos \theta_1 \cos \theta_2 ... \cos \theta_n (S_1 - S_3 + ...).$

Hence $\tan (\theta_1 + \theta_2 + ... + \theta_n) = \dfrac{S_1 - S_3 + ...}{1 - S_2 + S_4 + ...}.$

Example 1:

If α, β, γ are the roots of $x^3 + px^3 + qx + p = 0$, prove that $\tan^{-1}\alpha + \tan^{-1}\beta + \tan^{-1}\gamma = n\pi$, except when $q = 1$.

Solution:

Since α, β, γ are the roots of the given equation,

$S_1 = \alpha + \beta + \gamma = - p,\ S_2 = \alpha\beta + \beta\gamma + \gamma\alpha = q,\ S_3 = \alpha\beta\gamma = - p.$

If $\alpha = \tan \theta_1, \beta = \tan \theta_2, \gamma = \tan \theta_3$, then

$S_1 = \Sigma \tan \theta_1, \beta = \tan \theta_2, \gamma = \tan \theta_3$, then

$S_1 = \Sigma \tan \theta_1$, $S_2 = \Sigma \tan \theta_1 \tan \theta_2$, $S_3 = \tan \theta_1 \tan \theta_2 \tan \theta_3$.

Now $\tan(\theta_1 + \theta_2 + \theta_3) = \dfrac{S_1 - S_3}{1 - S_2} = \dfrac{-p-(-p)}{1-q} = \dfrac{0}{1-q}$,

$\Rightarrow \tan(\theta_1 + \theta_2 + \theta_3) = 0$, except when $q = 1$,

$\Rightarrow \tan(\theta_1 + \theta_2 + \theta_3) = \tan n\pi$, except when $q = 1$,

$\Rightarrow \theta_1 + \theta_2 + \theta_3 = np$, except when $q = 1$.

Hence $\tan^{-1}\alpha + \tan^{-1}\beta + \tan^{-1}\gamma = n\pi$, except when $q = 1$.

Example 2:

Prove that the equation $\cos 2\theta + p \cos\theta + q \sin\theta + r = 0$ *has, in general, four roots between 0 and* 2π*, and that the sum of the four roots is a multiple of* 2π.

Solution:

If $\tan\dfrac{\theta}{2} = t$, then $\cos\theta = \dfrac{1-t^2}{1+t^2}$, $\sin\theta = \dfrac{2t}{1+t^2}$.

Now $\cos 2\theta = 2\cos^2\theta - 1 = 2\left(\dfrac{1-t^2}{1+t^2}\right)^2 - 1 = \dfrac{1-6t^2+t^4}{(1+t^2)^2}$.

Substituting these values in the given equation, we get

$$\frac{1-6t^2+t^4}{(1+t^2)^2} + p\left(\frac{1-t^2}{1+t^2}\right) + q\left(\frac{2t}{1+t^2}\right) + r = 0$$

$\Rightarrow (1 - 6t^2 + t^4) + p(1 - t^2)(1 + t^2) + 2qt(1 + t^2) + r(1 + t^2)^2 = 0$

$\Rightarrow (1 - p + r)t^4\text{:}\ 2qt^3 + 2(r - 3)t^2 + 2qt + (1 + p + r) = 0$.

Let the roots of this equation be t_1, t_2, t_3 and t_4, and

$t_4 = \tan(\theta_i/2)$, $i = 1, 2, 3, 4$.

Then $S_1 = \Sigma \tan\dfrac{\theta}{2} = \Sigma t_1 = -\dfrac{2q}{1-p+r}$

and $S_3 = \Sigma t_1t_2t_2 = -\dfrac{2q}{1-p+r}$. Also $1 - S_2 + S_4 \neq 0$.

Now $\tan\dfrac{1}{2}(\theta_1 + \theta_2 + \theta_3 + \theta_4) = \dfrac{S_1 - S_3}{1 - S_2 + S_4} = 0$,

$= \tan n\pi$, n being any integer.

$\Rightarrow \dfrac{1}{2}(\theta_1 + \theta_2 + \theta_3 + \theta_4) = n\pi$. Hence $\theta_1 + \theta_2 + \theta_3 + \theta_4 = 2n\pi$.

Example 3:

If θ_1, θ_2, θ_3 be the three values of θ which satisfy the equation tan $2\theta = k \tan(\theta + \alpha)$ and such that no two of them differe by a multiple of π, show $\theta_1 + \theta_2 + \theta_3 + \alpha$ is a multiple of π.

Solution:

We have $\tan 2\theta = k \tan(\theta + \alpha)$

$$\Rightarrow \frac{2\tan\theta}{1-\tan^2\theta} = k.\frac{\tan\theta + \tan\alpha}{1-\tan\theta\tan\alpha}$$

$$\Rightarrow 2\tan\theta(1 - \tan\theta\tan\alpha) = k(\tan\theta + \tan\alpha)(1 - \tan^2\theta)$$

$$\Rightarrow 2\tan\theta - 2\tan^2\theta\tan\alpha = k(\tan\theta - \tan^3\theta + \tan\alpha - \tan\alpha\tan^2\theta)$$

$$\Rightarrow k\tan^3\theta + (k-2)\tan\alpha\tan^2\theta - (k-2)\tan\theta - k\tan\alpha = 0.$$

This equation being a cubic in $\tan\theta$ has three roots. Let the three roots be $\tan\theta_1$, $\tan\theta_2$, $\tan\theta_3$.

$$\therefore S_1 = \Sigma \tan\theta_1 = -\frac{k-2}{k}\tan\alpha.$$

$$S_2 = \Sigma \tan\theta_1\tan\theta_2 = -\frac{k-2}{k},$$

$$S_3 = \tan\theta_1\tan\theta_2\tan\theta_3 = \tan\alpha.$$

$$\text{Now } \tan(\theta_1 + \theta_2 + \theta_3) = \frac{S_1 - S_3}{1 - S_2} = \frac{\frac{1}{k}(2-k)\tan\alpha - \tan\alpha}{1 + \frac{1}{k}(k-2)}$$

$$= \frac{2\tan\alpha - k\tan\alpha - k\tan\alpha}{k + k - 2}$$

$$= \frac{2(k-1)\tan\alpha}{2(k-1)} = -\tan\alpha = \tan(n\pi - \alpha).$$

Thus $\theta_1 + \theta_2 + \theta_3 = n\pi - \alpha$, where n is any integer.

Hence $\theta_1 + \theta_2 + \theta_3 + \alpha = n\pi$, where n is any integer.

Example 4:

Prove that the equation

$$a^2\cos^2\theta + b^2\sin^2\theta + 2ga\cos\theta + 2fb\sin\theta + c = 0$$

has four roots and that the sum of the values of θ which satisfy it is an even multiple of π radians.

Solution:

Let $\tan\dfrac{\theta}{2} = t$. Then $\sin\theta = \dfrac{2t}{1+t^2}$, $\cos\theta = \dfrac{1-t^2}{1+t^2}$.

The given equation becomes

$$a^2\left(\frac{1-t^2}{1+t^2}\right)^2 + b^2\left(\frac{2t}{1+t^2}\right)^2 + 2ga\left(\frac{1-t^2}{1+t^2}\right) + 2fb\left(\frac{2t}{1+t^2}\right) + c = 0$$

$$\Rightarrow a^2(1-t^2)^2 + 4b^2t^2 + 2ga(1-t^2)(1+t^2) + 4fbt(1+t^2)^2 + c(1+t^2)^2 = 0$$

$$\Rightarrow t^4(c - 2ga + a^2) + 4fbt^2 + t^2(2c + 4b^2 - 2a^2) + 4bft + (c + 2ga + a^2) = 0.$$

This is a fourth degree equation in t and so it has four roots say t_1, t_2, t_3, t_4. Let $t_i = \tan(\theta_i/2)$, $i = 1, 2, 3, 4$.

Now $S_1 = \Sigma t_1 = \Sigma\tan\dfrac{\theta_1}{2} = \dfrac{-4fb}{c-2ga+a^2}$,

$S_3 = \Sigma t_1t_2t_3 = \dfrac{-4fb}{c-2ga+a^2}$. Also $1 - S_2 + S_4 \neq 0$.

$$\therefore \tan\frac{1}{2}(\theta_1 + \theta_2 + \theta_3 + \theta_4) = \frac{S_1 - S_3}{1 - S_2 + S_4} = 0 = \tan n\pi.$$

$$\Rightarrow \frac{1}{2}(\theta_1 + \theta_2 + \theta_3 + \theta_4) = n\pi,$$

Hence $\theta_1 + \theta_2 + \theta_3 + \theta_4 = 2n\pi$, where n is any integer.

Example 5:

Show that the equation $\cos(2\theta - \alpha) = p\cos(\theta - \beta)$ can be satisfied by four values θ_1, θ_2, θ_3, θ_4 of θ, of which no two differe by a multiple of 2π. Show that $\theta_1 + \theta_2 + \theta_3 + \theta_4 - 2\alpha = k\pi$, k being an even integer.

Solution:

We have $\cos(2\theta - \alpha) = p\cos(q - \beta)$

$\Rightarrow \cos[(\theta - \alpha/2) + (\theta - \alpha/2)] = p[\cos(\theta - \alpha/2) + (\alpha/2 - \beta)]$

$\Rightarrow \cos 2\theta = p\cos(\phi + c)$, where $\phi = \theta - \alpha/2$ and $c = \alpha/2 - \beta$

$\Rightarrow \cos^2\phi - \sin^2\phi = p[\cos c\cos\phi - \sin c\sin\phi]$

$$\Rightarrow \left(\frac{1-t^2}{1+t^2}\right) - \left(\frac{2t}{1+t^2}\right)^2 = p\cos c\left(\frac{1-t^2}{1+t^2}\right) - p\sin c\left(\frac{2t}{1+t^2}\right),\ t = \tan\phi/2.$$

$\Rightarrow (1 - t^2)^2 - 4t^2 = p \cos c\ (1 - t^2)(1 + t^2) - 2p \sin ct\ (1 + t^2)$

$\Rightarrow 1 + t^4 - 6t^2 = p \cos c\ (1 - t^4) - 2p \sin c\ (1 + t^3)$

$\Rightarrow (1 + p \cos c)\ t^4 + 2p \sin c\ t^3 - 6t^2 + 2p \sin ct + (1 - p \cos c) = 0.$

This is a fourth degree equation in t and so it has four values of t say t_1, t_2, t_3 and t_4. We have $t_i = \tan(\phi_i/2)$, i = 1, 2, 3, 4.

$$\therefore S_1 = \Sigma t_1 = \frac{-2p \sin c}{1 + p \cos c},$$

$$S_3 = \Sigma t_1t_2t_3 = \frac{-2p \sin c}{1 + p \cos c}, \text{ so that } S_1 - S_3 = 0.$$

$$\text{Now } \tan \frac{1}{2}(\phi_1 + \phi_2 + \phi_3 + \phi_4) = \frac{S_1 - S_3}{1 - S_1 + S_4} = 0, \ (1 - S_2 + S_4 \neq 0)$$

$$\therefore \frac{1}{2}(\phi_1 + \phi_2 + \phi_3 + \phi_4) = n\pi, \text{ n being any integer.}$$

$\Rightarrow \phi_1 + \phi_2 + \phi_3 + \phi_4 = 2n\pi$

$\Rightarrow (\theta_1 - \alpha/2) + (\theta_2 - \alpha/2) + (\theta_3 - \alpha/2) + (\theta_4 - \alpha/2) = k\pi, k = 2n$

$(\therefore \phi = \theta - \alpha/2)$

Hence $\theta + \theta_2 + \theta_3 + \theta_4 - 2\alpha = k\pi$, k being an even integer.

Example 6:

Show that there are four values of θ lying between 0 and 2π which satisfy the equation ap sec θ – bq cosec $\theta = c^2$, and that their sum is an odd multiple of 2π.

Solution:

We know

$$\cos \theta = \frac{1 - t^2}{1 + t^2} \sin \theta = \frac{2t}{1 + t^2}, \text{ when } \tan \frac{\theta}{2} = t.$$

The equation ap sec θ – bq cosec $\theta = c^2$ becomes

$$\frac{ap(1 + t^2)}{.1 - t^2} - \frac{bq(1 + t^2)}{2t} = c^2$$

$\Rightarrow bqt^4 + t^2 (2ap + 2c^2) = 0t^2 + t(2ap - 2c^2) - bq = 0.$

Since (1) is a fourth degree equation in t, so it has four values of t say t_1, t_2, t_3 and t_4. We have $t_i = \tan \theta/2$.

Then $S_1 = \Sigma t_1 = -(2ap + 2c^2)/bq$, $S_2 = \Sigma t_1t_2$

$S_3 = \Sigma\, t_1t_2t_3 = -(2ap - 2c^2)/bq,\ S_4 = t_1t_2t_3t_4 = -1.$

We know $\tan \frac{1}{2}(\theta_1 + \theta_2 + \theta_3 + \theta_4) = \dfrac{S_1 - S_3}{1 - S_2 + S_4}$

$\Rightarrow \cot \frac{1}{2}(\theta_1 + \theta_2 + \theta_3 + \theta_4) = \dfrac{1 - S_2 + S_4}{S_1 - S_3} = 0,\ (S_1 - S_3 \neq 0)$

$\Rightarrow \frac{1}{2}(\theta_1 + \theta_2 + \theta_3 + \theta_4) = (2n\pi + 1)\frac{\pi}{2}$, n being an integer

Hence $\theta_1 + \theta_2 + \theta_3 + \theta_4 = (2n + 1)\pi$, which is an odd multiple of 2p.

EXERCISES

1. Prove that $64(\cos^8\theta + \sin^8\theta) = \cos 8\theta + 28\cos 4\theta + 35$.
2. Prove that $\sin^9\theta\ 1/256[\sin^{-9}\theta \sin 7\theta + 35\sin 5\theta - 84$ in $3\sin\theta + 126\sin\theta]$.
3. Prove that $16\sin^8\Theta + \sin 5\Theta - \sin 3\theta - 10\sin\theta$.
4. Prove that $128\cos^8\theta = \cos 8\theta + 8\cos 6\theta + 28\cos 4 = 56\cos 2\theta + 35$.
5. Prove that $64(\cos^7\theta + 7\cos 7\theta + 7\cos 5\theta + 21\cos 3\theta + 35\cos\theta$.
6. Express $\sin^6\Theta\cos^2\Theta$ in termss of cosines of multiplie of Θ.

6

Gregory's Series and Trigonometrical Expansions

6.1 GEOGORY'S SERIES

To prove that, if θ lies within the closed interval $[-\pi/4, \pi/4]$, i.e., $-\pi/4 \le \theta \le \pi/4$, then

$$\theta = \tan\theta - \frac{1}{3}\tan^3\theta + \frac{1}{5}\tan^5\theta - \frac{1}{7}\tan^7\theta + \ldots ad. inf.$$

Proof: We have

$$(1 + i\tan\theta) = \left(1 + i\frac{\sin\theta}{\cos\theta}\right) = \frac{1}{\cos\theta}(\cos\theta + i\sin\theta)$$

$$= \sec\theta . e^{i\theta}.$$

Taking logarithim of both sides, we have

$\log(1 + i\tan\theta) = \log\sec\theta + \log e^{i\theta}$, (considering only principal values)

or $\log(1 + i\tan\theta) = \log\sec\theta + i\theta$.

Now since θ lies between $-\pi/4$ and $\pi/4$, $\tan\theta$ lies between -1 and 1, *i.e.,* $\tan\theta$ is numerically not greater than unity.

Therefore, $\log\sec\theta + i\theta = \log(1 + i\tan\theta)$

$$= i\tan\theta - \frac{i^2\tan^2\theta}{2} + \frac{i^3\tan^3\theta}{3} - \frac{i^4\tan^4\theta}{4} + \ldots, \qquad \ldots(1)$$

(expanding the R.H.S. by logarithmic series which is justified because $|i\tan\theta| = |\tan\theta| \le 1$)

$$= i\tan\theta + \frac{1}{2}\tan^2\theta - \frac{1}{3}i\tan^3\theta - \frac{1}{4}\tan^4\theta + \frac{1}{5}i\tan^5\theta + \ldots$$

Equating the imaginary parts on both sides, we have

$$\theta = \tan\theta - \frac{1}{3}\tan^3\theta + \frac{1}{5}\tan^5\theta - ... \qquad ...(2)$$

The general term on the R.H.S. is $\frac{(-1)^{n-1}}{2n-1}\tan^{2n-1}\theta$.

This expansion (2) is known as Gregory's series after the name of Jamess Gregory (1638-1675).

Another Form of Gregory's Series

In the series (2) if we put tan θ = x o that θ = $\tan^{-1}$ x, then we have another form of the Gregory's series as

$$\tan^{-1}x = x - \frac{x^3}{3} + \frac{x^5}{5} - \frac{x^7}{7} + ...\text{ad.inf. where } -1 \le x \le 1 \textit{ i.e., } |x| \le 1.$$

Cor. Equating real parts from both sides of (1), we have

$$\log\sec\theta = \frac{1}{2}\tan^2\theta - \frac{1}{4}\tan^4\theta + \frac{1}{6}\tan^6\theta - ...$$

6.2 GENERAL THEOREM OF GREGORY'S SERIES

If q lies between $n\pi \frac{1}{4}\pi$ *and* $n\pi + \frac{1}{4}\pi$, *both limits being inclusive, then*

$$\theta - n\pi = \tan\theta - \frac{1}{3}\tan^3\theta + \frac{1}{5}\tan^5\theta - ...ad.inf.$$

Proof: Let θ – nπ = ϕ *i.e.,* θ = nπ + ϕ. Then f lies between –π/4 and π/4.

Now 1 + i tan θ = 1 + i tan (nπ + ϕ) = 1 + i tan ϕ = sec ϕ (cos ϕ + i sin ϕ) = sec ϕ.$e^{i\phi}$.

Taking logarithm of both sides, we get

log (1 + i tan θ) = log sec ϕ + log $e^{i\phi}$

or log sec ϕ + iϕ = log (1 + i tan θ),

the expansion of which is valid because θ lies between $n\pi - \frac{1}{4}\pi$ and $n\pi + \frac{1}{4}\pi$ implies that tan θ is not numerically greater than 1.

$$\therefore \log\sec\phi + i\phi = i\tan\theta - \frac{1}{2}i^2\tan^2\theta + \frac{1}{3}i^3\tan^3\theta - \frac{1}{4}i^4\tan^4\theta + ...\infty$$

$$\left[\because \log(1+z) = z - \frac{1}{2}z^2 + \frac{1}{3}z^3 - ...\infty, \text{ if } |z| \le 1\right]$$

$$= i\tan\theta - \frac{1}{2}\tan^2\theta - \frac{1}{3}i\tan^3\theta - \frac{1}{4}\tan^4\theta + \ldots\infty.$$

Equating imaginary parts from both sides, we have

$$\phi = \tan\theta - \frac{1}{3}\tan^3\theta + \frac{1}{5}\tan^5\theta - \ldots\text{ad.inf.}$$

$$\therefore\ \theta - n\pi = \tan\theta - \frac{1}{3}\tan^3\theta + \frac{1}{5}\tan^5 - \ldots\text{ad.inf.}$$

6.3 VALUE OF p

The main use of Gregory's series is to find the value of π to various decimal places. Some mathematicians have designed different expressions based on Gregory's series for finding the value of π. Some important cases are given below:

(a) Gregory's Series **(Meerut, 1980)**

In the Gregory's series

$$\tan^{-1}x = x - \frac{1}{3}x^3 + \frac{1}{5}x^5 - \frac{1}{7}x^7 + \ldots\infty,$$

if we put $x = 1$, then we have

$$\tan^{-1}1 = 1 - \frac{1}{3} + \frac{1}{5} - \frac{1}{7} + \frac{1}{9} - \ldots\infty$$

or $$\frac{1}{4}\pi = 1 - \frac{1}{3} + \frac{1}{5} - \frac{1}{7} + \frac{1}{9} - \ldots\infty.$$

From this the value of π can be calculated. But this series does not converge rapidly and so a large number of terms will have to be taken in order to evaluate π correct to a certain decimal place. So several other series have been designed for this purpose.

(b) Euler's Series:

We have

$$\tan^{-1}\frac{1}{2} + \tan^{-1}\frac{1}{3} = \tan^{-1}\frac{\frac{1}{2}+\frac{1}{3}}{1-\frac{1}{6}} = \tan^{-1}\frac{\frac{5}{6}}{\frac{5}{6}} = \tan^{-1}1 = \frac{\pi}{4}$$

Thus $$\frac{\pi}{4} = \tan^{-1}\frac{1}{2} + \tan^{-1}\frac{1}{3}$$

$$= \left(\frac{1}{2} - \frac{1}{3}\cdot\frac{1}{2^3} + \frac{1}{5}\cdot\frac{1}{2^5} - \ldots\right) + \left(\frac{1}{3} - \frac{1}{3}\cdot\frac{1}{5}\cdot\frac{1}{3^5} - \ldots\right),$$

expanding both $\tan^{-1}\frac{1}{2}$ and $\tan^{-1}\frac{1}{3}$ by Gregory's series because $\frac{1}{2}<1$ and $\frac{1}{3}<1$

$$=\left(\frac{1}{2}+\frac{1}{3}\right)-\frac{1}{3}\left(\frac{1}{2^3}+\frac{1}{3^3}\right)+\frac{1}{5}\left(\frac{1}{2^5}+\frac{1}{3^5}\right)-\dots$$

From this the value of π can be calculated. This series is more rapidly convergent than the preceding one.

(c) Machin's Series:

We have

$$4\tan^{-1}\frac{1}{5}=2.2\tan^{-1}\frac{1}{5}=2\tan^{-1}\frac{\frac{2}{5}}{1-\left(\frac{1}{25}\right)}=2\tan^{-1}\frac{5}{12}$$

$$=\tan^{-1}\left\{\frac{2.\left(\frac{5}{12}\right)}{1-\left(\frac{5}{12}\right)^2}\right\}=\tan^{-1}\frac{120}{119}.$$

Now $4\tan^{-1}\frac{1}{5}-\frac{1}{4}\pi=\tan^{-1}\frac{120}{119}-\tan^{-1}1$ $\left[\because \frac{1}{4}\pi=\tan^{-1}1\right]$

$$=\tan^{-1}\frac{\frac{120}{119}-1}{1+\frac{120}{119}\times 1}=\tan^{-1}\frac{1}{239}.$$

Therefore, $\frac{\pi}{4}=4\tan^{-1}\frac{1}{5}-\tan^{-1}\frac{1}{239}$

or $$\frac{\pi}{4}=4\left[\frac{1}{5}-\frac{1}{5}\cdot\frac{1}{5^3}+\frac{1}{5}\cdot\frac{1}{5^5}-\dots\right]$$

$$-\left[\frac{1}{239}-\frac{1}{3}\cdot\frac{1}{(239)^3}+\frac{1}{5}\cdot\frac{1}{(239)^5}-\dots\right],$$

expanding both the terms on the R.H.S. by Gregory's series because $\frac{1}{5}<1$ and $\frac{1}{239}<1$.

The above is Machin's series anhd it is more rapidly convergent than Euler's series.

(d) Rutherford's Series

We have

$$\tan^{-1}\frac{1}{70}-\tan^{-1}\frac{1}{99}=\tan^{-1}\frac{\frac{1}{70}-\frac{1}{99}}{1+\frac{1}{70}\cdot\frac{1}{99}}=\tan^{-1}\frac{99-70}{70.99+1}$$

$$=\tan^{-1}\frac{29}{6931}=\tan^{-1}\frac{1}{239}$$

$$=4\tan^{-1}\frac{1}{5}-\frac{\pi}{4},\text{ as shown in Machin's series.}$$

$$\therefore\ \frac{\pi}{4}=4\tan^{-1}\frac{1}{5}-\tan^{-1}\frac{1}{70}+\tan^{-1}\frac{1}{99}$$

$$i.e.,\ \frac{\pi}{4}=4\left\{\frac{1}{5}-\frac{1}{3}\frac{1}{5^3}+\frac{1}{5}\cdot\frac{1}{5^5}-\ldots\right\}-\left\{\frac{1}{70}-\frac{1}{3}\cdot\frac{1}{(70)^3}+\frac{1}{5}\cdot\frac{1}{(70)^5}-\ldots\right\}$$

$$+\left\{\frac{1}{99}-\frac{1}{3}\cdot\frac{1}{(99)^3}+\frac{1}{5}\cdot\frac{1}{(99)^5}-\ldots\right\}$$

It is Rutherford's series. This is more convenient for expansion than Machin's series and converges equally rapidly.

SOLVED EXAMPLES

Example 1:

Assuming that $\theta - n\pi = \tan\theta - \frac{1}{3}\tan^3\theta + \frac{1}{5}\tan^5\theta - \ldots$, *when* θ *lies between*

$$\left(n\pi-\frac{1}{4}\pi\right) \text{ and } \left(n\pi+\frac{1}{4}\pi\right),$$

write down the value of n when θ *lies between*

(i) $\frac{7\pi}{4}$ *and* $\frac{9\pi}{4}$

(ii) $\frac{-3\pi}{4}$ *and* $\frac{-5\pi}{4}$

(iii) $\frac{11\pi}{4}$ *and* $\frac{13\pi}{4}$

(iv) $\frac{-15\pi}{4}$ *and* $\frac{-17\pi}{4}$

(v) $\frac{-19\pi}{4}$ and $\frac{21\pi}{4}$.

Solution:

(i) θ lies between $\left(\frac{7\pi}{4}\right)$ and $\left(\frac{9\pi}{4}\right)$

i.e., between $\left(2\pi-\frac{1}{4}\right)$ and $\left(2\pi+\frac{1}{4}\pi\right)$. $\therefore$ n = 2

(ii) θ lies between $-\left(\frac{3\pi}{4}\right)$ and $-\left(\frac{5\pi}{4}\right)$

i.e., between $\left(-\pi+\frac{1}{4}\pi\right)$ and $\left(-\pi-\frac{1}{4}\pi\right)$. $\therefore$ n = – 1.

(iii) θ lies between $\left(\frac{111\pi}{4}\right)$ and $\left(\frac{13\pi}{4}\right)$

i.e., between $\left(3\pi-\frac{1}{4}\pi\right)$ and $\left(3\pi+\frac{1}{4}\pi\right)$, $\therefore$ n = 3.

(iv) θ lies between $-\left(\frac{15\pi}{4}\right)$ and $\left(-\frac{17\pi}{4}\right)$

i.e., between $\left(-4\pi+\frac{1}{4}\pi\right)$ and $\left(-4\pi-\frac{1}{4}\pi\right)$. $\therefore$ n = 4.

(v) θ lies between $\left(\frac{19\pi}{4}\right)$ and $\left(\frac{21\pi}{4}\right)$

i.e., between $\left(5\pi-\frac{1}{4}\pi\right)$ and $\left(5\pi+\frac{1}{4}\pi\right)$. $\therefore$ n = 5.

Example 2:

If f lies between $\left(\frac{\pi}{4}\right)$ *and* $\left(\frac{3\pi}{4}\right)$, *show that*

$$\phi=\frac{1}{2}\pi-\cot\phi+\frac{1}{3}\cot^3\phi-\frac{1}{5}\cot^5\phi+\ldots\infty.$$ **(Kanpur, 1991)**

Solution:

We have to prove that $\frac{1}{2}\pi-\phi=\cot\phi-\frac{1}{3}\cot^3\phi+\frac{1}{5}\cot^5\phi-\ldots\infty$.

Let $\frac{1}{2}\pi-\phi=\theta$. When $\phi=\left(\frac{\pi}{4}\right)$, we have $\theta=\frac{1}{2}\pi-\frac{1}{4}\pi=\frac{1}{4}\pi$ and when

$\phi = \left(\frac{3\pi}{4}\right)$, we have $\theta = \frac{1}{2}\pi - \frac{3}{4}\pi = -\frac{1}{4}\pi$. Thus if ϕ lies between $\left(\frac{\pi}{4}\right)$ and $\left(\frac{3\pi}{4}\right)$, then θ lies between $\left(\frac{-\pi}{4}\right)$ and $\left(\frac{\pi}{4}\right)$ *i.e.,* for θ the conditions of Gregory's series are satisfied.

Therefore we have

$$\theta = \tan\theta - \frac{1}{3}\tan^3\theta + \frac{1}{5}\tan^5\theta - \ldots\infty, \qquad \text{(by Gregory's series)}$$

i.e., $$\frac{1}{2}\pi - \phi = \tan\left(\frac{1}{2}\pi - \phi\right) - \frac{1}{3}\tan^3\left(\frac{1}{2}\pi - \phi\right) + \frac{1}{5}\tan^5\left(\frac{1}{2}\pi - \phi\right) - \ldots\infty.$$

$$\left[\because \theta = \frac{1}{2}\pi - \phi\right]$$

$$= \cot\phi - \frac{1}{3}\cot^3 + \frac{1}{5}\cot^5\phi - \ldots\infty.$$

This proves the result.

Example 3:

Sun the series $\frac{1}{2^3} - \frac{1}{3.2^7} + \frac{1}{5.2^{11}} - \ldots ad.inf$.

Solution:

We have

$$\frac{1}{2^3} - \frac{1}{3.2^7} + \frac{1}{5.2^{11}} + \ldots = \frac{1}{2}\left[\frac{1}{2^2} - \frac{1}{3.2^6} + \frac{1}{5.2^{10}} - \ldots\right],$$

$$\left(\text{taking } \frac{1}{2} \text{ common}\right)$$

$$= \frac{1}{2}\left[\frac{1}{(2^2)} - \frac{1}{3(2^2)^3} + \ldots\right] = \frac{1}{2}\tan^{-1}\left(\frac{1}{2^2}\right),$$

by Gregory's series because $\frac{1}{2^2} < 1$.

$$\left[\text{Note that by Gregory's series } \tan^{-1}x = x - \frac{1}{3}x^3 + \frac{1}{5}x^5 - \ldots\infty, \text{ if } |x| \le 1\right]$$

$$= \frac{1}{2}\tan^{-1}\frac{1}{4}.$$

Example 4:

Sum to infinity the series

(i) $1-\dfrac{1}{3.4^2}+\dfrac{1}{5.4^4}-\ldots ad.inf.;$

(ii) $1-\dfrac{1}{3^2}+\dfrac{1}{5.3^2}-\dfrac{1}{7.3^3}+\ldots ad.inf.$

Solution:

(i) The given series

$$1-\frac{1}{3.4^2}+\frac{1}{5.4^4}-\ldots\text{ad.inf.}$$

$$=4\left[\frac{1}{4}-\frac{1}{3.4^3}+\frac{1}{5.4^5}-\ldots\text{ad.inf.}\right],\qquad\text{(taking 4 common)}$$

$$=4\tan^{-1}\frac{1}{4},\ \text{by Gregory's series because } \frac{1}{4}<1.$$

(ii) The given series

$$1-\frac{1}{3^2}+\frac{1}{5.3^2}-\frac{1}{7.3^3}+\ldots=1-\frac{1}{3.3}+\frac{1}{5.3^2}-\frac{1}{7.3^3}+\ldots$$

$$=\sqrt{3}\left[\frac{1}{\sqrt{3}}-\frac{1}{3.\left(\sqrt{3}\right)^3}+\frac{1}{5\left(\sqrt{3}\right)^5}-\frac{1}{7.\left(\sqrt{3}\right)^7}+\ldots\right]$$

$$=\sqrt{3}\left\{\tan^{-1}\left(\frac{1}{\sqrt{3}}\right)\right\},\ \text{by Gregory's series because } \frac{1}{\sqrt{3}}<1$$

$$=\sqrt{3}.\left(\frac{\pi}{6}\right)=\frac{\left(\frac{\pi}{\sqrt{3}}\right)}{6}.$$

Note: This question can also be written as

Prove that $\pi=2\sqrt{3}\left\{1-\dfrac{1}{3^2}+\dfrac{1}{5.3^2}-\dfrac{1}{7.3^3}+\ldots\right\}$

(Garhwal, 1983; Kanpur, 91; Meerut, 88)

Example 5:

Prove that

$$\frac{\pi}{4}=\frac{17}{21}-\frac{713}{81\times343}+\ldots+\frac{(-1)^{n+1}}{2n-1}\left\{\frac{2}{3}9^{1-n}+7^{1-2n}\right\}+\ldots\qquad\textbf{(Kanpur, 1992)}$$

Solution:

The nth term of the given series on the R.H.S. is

$$T_n = \frac{(-1)^{n+1}}{2n-1}\left\{\frac{2}{3}\cdot\left(3^2\right)^{1-n} + 7^{1-2n}\right\}$$

$$= \frac{(-1)^{n+1}}{2n-1}\left\{\frac{2}{3}\cdot 3^{2-2n} + 7^{1-2n}\right\} = \frac{(-1)^{n+1}}{(2n-1)}\left[\frac{2}{3^{2n-1}} + \frac{1}{7^{2n-1}}\right].$$

Putting n = 1, 2, 3, ..., etc., we have

first term $= T_1 = \left(\frac{2}{3} + \frac{1}{7}\right);$

second term $= T_2 = -\frac{1}{3}\left(\frac{2}{3^3} + \frac{1}{7^3}\right) = -\frac{2}{3.3^3} - \frac{1}{3.7^3};$

third term $= T_3 = \frac{1}{5}\left(\frac{2}{3^5} + \frac{1}{7^5}\right) = \frac{2}{5}\cdot\frac{1}{3^5} + \frac{1}{5.7^5};$

and so on.

$\therefore$ the sum of the series on the R.H.S.

$= T_1 + T_2 + T_3 + \ldots$ ad.inf.

$$= \left(\frac{2}{3} + \frac{1}{7}\right) - \left(\frac{2}{3.3^3} + \frac{1}{3.7^3}\right) + \left(\frac{2}{5.3^5} + \frac{1}{5.7^5}\right) - \ldots$$

$$= 2\left(\frac{1}{3} - \frac{1}{3.3^3} + \frac{1}{5.3^5} - \ldots\right) + \left(\frac{1}{7} - \frac{1}{3.7^3} + \frac{1}{5.7^5} - \ldots\right)$$

$$= 2\tan^{-1}\left(\frac{1}{3}\right) + \tan^{-1}\left(\frac{1}{7}\right), \text{ (by Gregory's series)}$$

$$= \tan^{-1}\left\{\frac{\frac{1}{3} + \frac{1}{3}}{1 - \frac{1}{3}\cdot\frac{1}{3}}\right\} + \tan^{-1}\frac{1}{7} = \tan^{-1}\frac{3}{4} + \tan^{-1}\frac{1}{7}$$

$$= \tan^{-1}\left\{\frac{\frac{3}{4} + \frac{1}{7}}{1 - \frac{3}{4}\cdot\frac{1}{7}}\right\} = \tan^{-1} 1 = \frac{\pi}{4}.$$

Example 6:

Express $\tan^{-1}(\cos\theta + i\sin\theta)$ *in the form* $A + iB$ *and deduce that*

(i) $\cos\theta - \frac{1}{3}\cos 3\theta + \frac{1}{5}\cos 5\theta - ... = \pm\frac{1}{4}\pi$; and

(ii) $\sin\theta - \frac{1}{3}\sin 3\theta + \frac{1}{5}\sin 5\theta - ... = \frac{1}{2}\log\left\{\pm\tan\left(\frac{1}{4}\pi + \frac{1}{2}\theta\right)\right\}$.

Solution:

Let $\tan^{-1}(\cos\theta + i\sin\theta) = A + iB$

so that $\tan^{-1}(\cos\theta - i\sin\theta) = A - iB$, (complex conjugates).

$\therefore\ 2A = (A + iB) + (A - iB)$

$= \tan^{-1}(\cos\theta + i\sin\theta) + \tan^{-1}(\cos\theta - i\sin\theta)$

$$= \tan^{-1}\left\{\frac{(\cos\theta + i\sin\theta) + (\cos\theta - i\sin\theta)}{1 - (\cos\theta + i\sin\theta)(\cos\theta - i\sin\theta)}\right\}$$

$$= \tan^{-1}\left\{\frac{2\cos\theta}{1 - (\cos^2\theta + \sin^2\theta)}\right\} = \tan^{-1}\left(\frac{2\cos\theta}{1-1}\right) = \tan^{-1}\left(\frac{2\cos\theta}{0}\right)$$

$= \tan^{-1}(\pm\infty) = \pm\frac{1}{2}\pi$. [$\because$ θ lies between $-\pi$ and π]

Therefore, $A = \pm\frac{1}{4}\pi$. ...(1)

Again $2iB = (A + iB) - (A - iB) = \tan^{-1}(\cos\theta + i\sin\theta)$
$- \tan^{-1}(\cos\theta - i\sin\theta)$

$$= \tan^{-1}\left\{\frac{(\cos\theta + i\sin\theta) - (\cos\theta - i\sin\theta)}{1 + (\cos\theta + i\sin\theta)(\cos\theta - i\sin\theta)}\right\}$$

$$= \tan^{-1}\left\{\frac{2i\sin\theta}{1 + (\cos^2\theta + \sin^2\theta)}\right\}$$

$$= \tan^{-1}\left\{\frac{2i\sin\theta}{1+1}\right\} = \tan^{-1}(i\sin\theta)$$

or $\tan(2iB) = i\sin\theta$ or $\tanh 2B = \sin\theta$

or $\frac{e^{2B} - e^{-2B}}{e^{2B} + e^{-2B}} = \frac{\sin\theta}{1}$ or $\frac{2e^{2B}}{2e^{-2B}} = \frac{1+\sin\theta}{1-\sin\theta}$

[By componendo and dividendo]

i.e., $$e^{4B} = \frac{1+\sin\theta}{1-\sin\theta} = \frac{\left\{\cos\left(\frac{\theta}{2}\right)+\sin\left(\frac{\theta}{2}\right)\right\}^2}{\left\{\cos\left(\frac{\theta}{2}\right)-\sin\left(\frac{\theta}{2}\right)\right\}} = \pm\tan\left(\frac{\pi}{4}+\frac{\theta}{2}\right).$$

$$\therefore \quad B = \frac{1}{2}\log\left\{\pm\tan\left(\frac{\pi}{4}+\frac{\theta}{2}\right)\right\} \qquad ...(2)$$

Now $\tan^{-1}(\cos\theta + i\sin\theta) = A + iB$

or $\tan^{-1}(e^{i\theta}) = A + iB$

or $e^{i\theta} - \frac{1}{3}e^{3i\theta} + \frac{1}{5}e^{5i\theta} - ... = A + iB$, using Gregory's series

or $(\cos\theta + i\sin\theta) - \frac{1}{3}(\cos 3\theta + i\sin 3\theta) + \frac{1}{5}(\cos 5\theta + i\sin 5\theta) - ... = A + iB$

or $\left(\cos\theta - \frac{1}{3}\cos 3\theta + \frac{1}{5}\cos 5\theta - ...\right) + i\left(\sin\theta - \frac{1}{3}\sin 3\theta + \frac{1}{5}\sin 5\theta - ...\right)$

$= A + iB = \pm\frac{1}{4}\pi + \frac{1}{2}i\log\left\{\pm\tan\left(\frac{1}{4}\pi + \frac{1}{2}\theta\right)\right\}$, by (1) and (2)

Equating real and and imaginary parts, we have

$$\cos\theta - \frac{1}{3}\cos 3\theta + \frac{1}{5}\cos 5\theta - ... = \pm\frac{1}{4}$$

and $\sin\theta - \frac{1}{3}\sin 3\theta + \frac{1}{5}\sin 5\theta - ... = \frac{1}{2}\log\left\{\pm\tan\left(\frac{1}{4}\pi + \frac{1}{2}\theta\right)\right\}$.

Example 8:

Expand $\frac{1 + a\cos\theta}{1 + 2a\cos\theta + a^2}$ *in a series of cosines of multiples of* θ, *where* $a < 1$.

Solution:

We have

$$\frac{1 + a\cos\theta}{1 + 2a\cos\theta + a^2} = \frac{1 + \frac{1}{2}a\left(e^{i\theta} + e^{-i\theta}\right)}{1 + a\left(e^{i\theta} + e^{-i\theta}\right) + a^2} \qquad \left[\because \cos\theta = \frac{1}{2}\left(e^{i\theta} + e^{-i\theta}\right)\right]$$

$$= \frac{1}{2}\cdot\frac{2 + a\left(e^{i\theta} + e^{-i\theta}\right)}{1 + a\left(e^{i\theta} + e^{-i\theta}\right) + a^2.e^{i\theta}.e^{-i\theta}}$$

$$= \frac{1}{2} \frac{\left(1+a e^{i\theta}\right)+\left(1+a e^{-i\theta}\right)}{\left(1+a e^{i\theta}\right)\left(1+a e^{-i\theta}\right)} = \frac{1}{2}\left[\frac{1}{\left(1+a e^{i\theta}\right)}+\frac{1}{\left(1+a e^{-i\theta}\right)}\right]$$

$$= \frac{1}{2}\left[\left(1+a e^{i\theta}\right)^{-1}+\left(1+a e^{-i\theta}\right)^{-1}\right]$$

$$= \frac{1}{2}[(1 - a\ e^{i\theta} + a^2\ a^{2i\theta} - a^3\ e^{3i\theta} + ...)$$

$$+ (1 - a\ e^{-i\theta} + a^2\ e^{-2i\theta} - a^3\ e^{-3i\theta} + ...)],$$

expanding by binomial theorem which is justified since $a < 1$

$$= 11 - a\left(\frac{e^{i\theta}+e^{-i\theta}}{2}\right)+a^2\left(\frac{e^{2i\theta}+e^{-2i\theta}}{2}\right)-a^3\left(\frac{e^{3i\theta}+e^{-3i\theta}}{2}\right)+...$$

$$= 1 - a \cos\theta + a^2 \cos 2\theta - a^3 \cos 3\theta + ...$$

Example 8:

Expand $\dfrac{\cos\theta - a\cos(\theta-\phi)}{1-2a\cos\phi+a^2}$ *in an infinite series if* $a < 1$.

Solution:

We have

$$\frac{\cos\theta - a\cos(\theta-\phi)}{1-2a\cos\phi+a^2} = \frac{1}{2}\cdot\frac{\left(e^{i\theta}+e^{-i\theta}\right)-a\left\{e^{i(\theta-\phi)}+e^{-i(\theta-\phi)}\right\}}{1-a\left(e^{i\phi}+e^{-i\phi}\right)+a^2}$$

$$= \frac{1}{2}\cdot\frac{e^{i\theta}\left(1-a e^{-i\phi}\right)+e^{-i\theta}\left(1-a e^{i\phi}\right)}{\left(1-a e^{-i\phi}\right)\left(1-a e^{i\phi}\right)}$$

$$= \frac{1}{2}\cdot\left[e^{i\theta}\left(1-a e^{i\phi}\right)^{-1}+e^{-i\theta}\left(1-a e^{-i\phi}\right)^{-1}\right]$$

$$= \frac{1}{2}[e^{i\theta}\ (1 + a\ e^{i\phi} + a^2\ e^{2i\phi} + a^3\ e^{3i\phi} + ...)$$

$$+ e^{-i\theta}\ (1 + a\ e^{i\phi} + a^2\ e^{-2i\phi} + a^3\ e^{-3i\phi} + ...)].$$

on expanding by binomial theorem which is valid because $a < 1$

$$= \frac{1}{2}\left[e^{i\theta}+e^{-i\theta}\right]+\frac{1}{2}a\left[e^{i(\theta+\phi)}+e^{-i(\theta+\phi)}\right]+\frac{1}{2}a^2\left[e^{i(\theta+2\phi)}+e^{-i(\theta+2\phi)}\right]+...\infty$$

$$= \cos\theta + a\cos(\theta+\phi) + a^2\cos(\theta+2\phi) + a^3\cos(\theta+3\phi) + ...\ \infty.$$

Example 9:

Expand e^{ax} sin bx in a series of ascending powers of x. Also find the coefficient of x^n in this expansion.

Solution:

We have e^{ax} sin bx

$$= \frac{1}{2i} e^{ax}\left[e^{ibx} - e^{-ibx}\right] = \frac{1}{2i}\left[e^{(a+ib)x} - e^{(a-ib)x}\right]$$

$$= \frac{1}{2i}\left[\left\{1 + (a+ib)x + \frac{(a+ib)^2 x^2}{2!} + \ldots\right\} - \left\{a + (a-ib)x + \frac{(a-ib)^2 x^2}{2!} + \ldots\right\}\right]$$

[$\because e^x = 1 + x/1! + x^2/2! + x^3/3! + \ldots$]

$$= \frac{1}{2i}\left[x\{(a+ib) - (a-ib)\} + \frac{x^2}{2!}\left\{(a+ib)^2 - (a-ib)^2\right\} + \ldots\right] \qquad \ldots(1)$$

Now put a = r cos θ and b = r sin θ, so that

$r = \sqrt{(a^2 + b^2)}$ and $\theta = \tan^{-1}\left(\frac{b}{a}\right)$.

Then the coefficient of x^n in (1)

$$= \frac{1}{2i}\frac{1}{n!}\left[(a+ib)^n - (a-ib)^n\right] = \frac{1}{2i(n!)}$$

$$[(r\cos\theta + ir\sin\theta)^n - (r\cos\theta - ir\sin\theta)^n]$$

$$= \frac{r^n}{2i(n!)}\left[(\cos n\theta + i\sin n\theta) - (\cos n\theta - i\sin n\theta)\right]$$

$$= \frac{r^n}{2i(n!)} 2i \sin n\theta = \frac{r^n}{n!}\sin n\theta.$$

∴ from (1), we have

$$e^{ax}\sin bx = x(r\sin\theta) + \frac{x^2 r^2 \sin 2\theta}{2!} + \frac{x^3 r^3 \sin 3\theta}{3!} + \ldots + \frac{x^n r^n \sin n\theta}{n!} + \ldots$$

Also the coefficient of x^n

$$= \frac{r^n}{n!}\sin n\theta = \frac{(a^2 + b^2)^{n/2}}{n!}\sin\left\{n\tan^{-1}\frac{b}{a}\right\}.$$

The expansion is valid for all values of a, b, x.

Example 10:

Expand $e^{a\cos\phi}\cos(\theta + a\sin\phi)$ in an infinite series.

Solution:

We have

$e^{a\cos\phi}\cos(\theta + a\sin\phi)$

$$= e^{a\cos\phi}.\left[\frac{e^{i(\theta+a\sin\phi)} + e^{-i(\theta+a\sin\phi)}}{2}\right] \quad \textbf{(Note)}$$

$$= \frac{1}{2}\left\{e^{i\theta}.e^{a(\cos\phi+i\sin\phi)} + e^{-i\theta}.e^{a(\cos\phi-i\sin\phi)}\right\}$$

$$= \frac{1}{2}\left\{e^{i\theta}.e^{ae^{i\phi}} + e^{-i\theta}.e^{ae^{-i\phi}}\right\} \quad (\because \cos\phi + i\sin\phi = e^{i\phi})$$

$$= \frac{1}{2}\left[e^{i\theta}\left(1 + ae^{i\phi} + \frac{1}{2!}a^2e^{2i\phi} + \frac{1}{3!}a^3e^{3i\phi} + \ldots\right)\right.$$

$$\left. + e^{-i\theta}\left(1 + ae^{-i\phi} + \frac{1}{2!}^2 e^{-2i\phi} + \frac{1}{3!}a^3e^{-3i\phi} + \ldots\right)\right]$$

$$= \frac{1}{2}\left[\left\{e^{i\theta} + e^{-i\theta}\right\} + a\left\{e^{i(\theta+\phi)}\right\} + e^{-i(\theta+\phi)}\right.$$

$$\left. + \frac{1}{2!}a^2\left\{e^{i(\theta+2\phi)} + e^{-i(\theta+2\phi)}\right\} + \frac{1}{3!}a^3\left\{e^{i(\theta+3\phi)} + e^{-i(\theta+3\phi)} + \ldots\right\}\right]$$

$= \cos\theta + a\cos(\theta + \phi) + (1/2!)\,a^2\cos(\theta + 2\phi) + (1/3!)\,a^3\cos(\theta + 3\phi) + \ldots$

Example 11:

Expand log cos $\left(\frac{1}{4}\pi + \theta\right)$ in a series of sine and cosine of multiples of θ.

Solution:

We have

$$\log\cos\left(\frac{1}{4}\pi + \theta\right) = \frac{1}{2}\log\cos^2\left(\frac{1}{4}\pi + \theta\right)$$

$$= \frac{1}{2}\log\frac{2\cos^2\left(\frac{1}{4}\pi + \theta\right)}{2} = \frac{1}{2}\log\frac{1 + \cos\left(\frac{1}{2}\pi + 2\theta\right)}{2}$$

$$= \frac{1}{2}\log\frac{1-\sin 2\theta}{2} = \frac{1}{2}\log\left\{\frac{1-\left(\frac{1}{2i}\right)\left(e^{2i\theta}-e^{-2i\theta}\right)}{2}\right\}$$

$$= \frac{1}{2}\log\left\{\frac{2-\left(\frac{i}{i^2}\right)\left(e^{2i\theta}-e^{-2i\theta}\right)}{4}\right\} = \frac{1}{2}\log\frac{2+i\left(e^{2i\theta}-e^{-2i\theta}\right)}{4}$$

$$= \frac{1}{2}\log\frac{\left(1+ie^{2i\theta}\right)\left(1-ie^{-2i\theta}\right)}{2^2} \qquad \textbf{(Note)}$$

$$= \frac{1}{2}\log\left\{\left(1+ie^{2i\theta}\right)\left(1+ie^{2i\theta}\right)\left(1-ie^{-2i\theta}\right)\right\} - \frac{1}{2}.2\log 2$$

$$= \frac{1}{2}\left[\log\left(1+ie^{2^{i\theta}}\right)+\log\left(1-ie^{-2i\theta}\right)\right]-\log 2$$

$$= \frac{1}{2}\left[\left(ie^{2i\theta}-\frac{1}{2}.i^2e^{4i\theta}+\frac{1}{3}.i^3e^{6i\theta}-\ldots\right)\right.$$

$$\left.-\left(ie^{-2i\theta}+\frac{1}{2}i^2e^{-4i\theta}+\frac{1}{3}i^3e^{-6i\theta}+\ldots\right)\right]-\log 2,$$

on expanding by logarithmic expansion

$$= \left[i\cdot\left(\frac{e^{2i\theta}-e^{-2i\theta}}{2}\right)+\frac{1}{2}\cdot\left(\frac{e^{4i\theta}+e^{-4i\theta}}{2}\right)-\frac{1}{3}i\cdot\left(\frac{e^{6i\theta}-e^{-6i\theta}}{2}\right)+\ldots\right]-\log 2$$

$$= \left[i.i\sin 2\theta+\frac{1}{2}.\cos 4\theta-\frac{1}{3}i.i\sin 6\theta+\ldots\right]-\log 2$$

$$= -\sin 2\theta+\frac{1}{2}\cos 4\theta+\frac{1}{3}\sin 6\theta+\ldots-\log 2$$

$$= -\log 2-\sin 2\theta+\frac{1}{2}\cos 4\theta+\frac{1}{3}\sin 6\theta-\frac{1}{4}\cos 8\theta\ldots\text{ad.inf.}$$

Example 12:

Prove that

$$\log\sin\theta = -\log 2-\cos 2\theta-\frac{1}{2}\cos 4\theta-\frac{1}{3}\cos 6\theta-\ldots\text{ ad.inf.}$$

Hence prove that if a and b the sides of a plane triangle, A and B the opposite angles,

$$log\ b - log\ a = (cos\ 2A - cos\ 2B) + \frac{1}{2} cos\ 4\theta - \frac{1}{3}\ co\ 6\theta - ...\ ad.inf.$$

Solution:

We have

$$\log \sin\theta = \frac{1}{2}\log\sin^2\theta = \frac{1}{2}\log\left\{\frac{(1-\cos 2\theta)}{2}\right\} = \frac{1}{2}\log\left\{\frac{1-\frac{1}{2}\left(e^{2i\theta}+e^{-2i\theta}\right)}{2}\right\}$$

$$= \frac{1}{2}\log\left\{\frac{2-\left(e^{2i\theta}+e^{-2i\theta}\right)}{4}\right\} = \frac{1}{2}\log\left\{\frac{\left(1-e^{2i\theta}\right)\left(1-e^{-2i\theta}\right)}{4}\right\}$$

$$= \frac{1}{2}\log\left\{\left(1-e^{2i\theta}\right)\left(1-e^{-2i\theta}\right)\right\} - \frac{1}{2}\log 4$$

$$= \frac{1}{2}\left[\log\left(1-e^{2i\theta}\right)+\log\left(1-e^{-2i\theta}\right)\right] - \log 2$$

$$= \frac{1}{2}\left[\left(-e^{2i\theta}-\frac{1}{2}e^{4i\theta}-...\right)+\left(-e^{-2i\theta}-\frac{1}{2}e^{-4i\theta}\right)-...\right] - \log 2$$

$$= -\frac{1}{2}\left[\left(e^{2i\theta}+e^{-2i\theta}\right)+\frac{1}{2}\left(e^{4i\theta}+e^{-4i\theta}\right)+...\right] - \log 2$$

$$= \left[-\cos 2\theta - \frac{1}{2}\cos 4\theta - \frac{1}{3}\cos 6\theta - ...\right] - \log 2$$

$$= -\log 2 - \cos 2\theta - \frac{1}{2}\cos 4\theta - \frac{1}{3}\cos 6\theta - ...\text{ad.inf.} \qquad ...(1)$$

This proves the first result.

Also we know that

$$\frac{\sin A}{a} = \frac{\sin B}{b} = k(\text{say}).$$

$\therefore$ ak = sin A, and bk = sin B.

Using (1), we have

$$\log ak = \log \sin A = -\log 2 - \cos 2A - \frac{1}{2}\cos 4A - ... \qquad ...(2)$$

$$\text{and } \log bk = \log \sin B = -\log 2 - \cos 2B - \frac{1}{2}\cos 4B - ...$$

Subtracting (2) and (3), we have

$$\log bk - \log ak = (\cos 2A - \cos 2B) + \frac{1}{2}(\cos 4A - co\ 4B) + ...$$

or (log b + log k) – (log a + log k) = (cos 2A – cos 2B)

$$+ \frac{1}{2} (\cos 4A - \cos 4B) + \ldots$$

or log b – log a = (cos 2A – cos 2B) + $\frac{1}{2}$ (cos 4A – cos 4B) + ...

This proves the second result.

Example 13:

In any triangle where a > b, prove that

$$\log c = \log a - \frac{b}{a}\cos C - \frac{1}{2}\frac{b^2}{a^2}\cos 2C - \frac{1}{3}\frac{b^3}{a^3}\cos 3C - \ldots$$

Solution:

We know that

$$c^2 = a^2 + b^2 - 2ab \cos C = a^2 \left[1 + \frac{b^2}{a^2} - 2\frac{b}{a}\cos C\right]$$

$= a^2 [1 + x^2 - 2x \cos C]$, putting $\left(\frac{b}{a}\right) = x$

$$= a^2\left[1 + x^2 - 2x\cdot\frac{1}{2}\left(e^{iC} + e^{-iC}\right)\right] = a^2\left\{\left(1 - xe^{iC}\right)\left(1 - xe^{-iC}\right)\right\}.$$

Taking logarithm of both sides, we have

$2 \log c = 2 \log a + \log (1 - xe^{iC}) + \log (1 - xe^{-iC})$

$$= 2 \log a + \left(-xe^{iC} - \frac{1}{2}x^2e^{2iC} - \frac{1}{3}x^3e^{3iC} - \ldots\right)$$

$$+\left(-xe^{-iC} - \frac{1}{2}x^2e^{-2iC} - \frac{1}{3}x^3e^{-3iC} - \ldots\right),$$

expanding by logarithmic expansion since x < 1, a being given to be greater than b

$$= 2\log a - x\left(e^{iC} + e^{-iC}\right) - \frac{1}{2}x^2\left(e^{2iC} + e^{-2iC}\right) - \frac{1}{3}x^3\left(e^{3iC} - e^{-3iC}\right) - \ldots$$

$$= 2\log a - 2x\cos C - \frac{1}{2}x^2.2\cos 2C - \frac{1}{3}x^3.2\cos 3C - \ldots$$

$$= \log c - \log a - \frac{b}{a}\cos C - \frac{1}{2}\frac{b^2}{a^2}\cos 2C - \frac{1}{3}\frac{b^3}{a^3}\cos 3C - \ldots$$

[Putting x = b/a]

Example 14:

If $\tan\theta = \dfrac{a\sin\alpha}{1+a\cos\alpha}$, *prove that* $\theta = a\sin\alpha - \dfrac{1}{2}a^2\sin 2\alpha + \dfrac{1}{3}a^3\sin 3\alpha - \ldots$

Solution:

We have $\tan\theta = \dfrac{a\sin\alpha}{1+a\cos\alpha}$

$$\therefore \quad \frac{\sin\theta}{\cos\theta} = \frac{a\sin\alpha}{1+a\cos\alpha}$$

or $\quad \dfrac{i\sin\theta}{\cos\theta} = \dfrac{ia\sin\alpha}{1+a\cos\alpha}$, multiplying both sides by i.

Applying componendo and dividendo, we have

$$\frac{\cos\theta + i\sin\theta}{\cos\theta - i\sin\theta} = \frac{1+a\cos\alpha + ia\sin\alpha}{1+a\cos\alpha - ia\sin\alpha}$$

or $\quad \dfrac{e^{i\theta}}{e^{-i\theta}} = \dfrac{1+ae^{i\alpha}}{1+ae^{-i\alpha}}$ or $e^{2i\theta} = \dfrac{1+ae^{i\alpha}}{1+ae^{-i\alpha}}$.

Taking logarithm of both sides, we have

$2i\theta = \log(1 + ae^{i\alpha}) - \log(1 + ae^{-i\alpha})$

$$= \left(ae^{i\alpha} - \frac{1}{2}a^2e^{i2\alpha} + \frac{1}{3}a^3e^{i3\alpha} - \ldots\right) - \left(ae^{-i\alpha} - \frac{1}{2}a^2e^{-2\alpha} + \frac{1}{3}a^3e^{-3i\alpha} - \ldots\right),$$

the logarithmic expansions are valid if $|a| < 1$

$$= a\left(e^{i\alpha} - e^{-i\alpha}\right) - \frac{1}{2}a^2\left(e^{i2\alpha} - e^{-2\alpha}\right) + \frac{1}{3}a^3\left(e^{i3\alpha} - e^{-i3\alpha}\right) - \ldots$$

$$\therefore \quad \theta = a\left(\frac{e^{i\alpha} - e^{-i\alpha}}{2i}\right) - \frac{1}{2}a^2\frac{e^{i2\alpha} - e^{-i2\alpha}}{2i} + \frac{1}{3}a^3\frac{e^{i3\alpha} - e^{-i3\alpha}}{2i} - \ldots$$

$$= a\sin\alpha - \frac{1}{2}a^2\sin 2\alpha + \frac{1}{3}a^3\sin 3\alpha - \ldots$$

Example 15:

If $\tan\theta = x + \tan\alpha$, *show that*

$$\theta = \alpha + x\cos^2\alpha - \frac{1}{2}x^2\cos^2\alpha\sin 2\alpha - \frac{1}{3}x^3\cos^3\alpha\cos 3\alpha + \ldots$$

Solution:

We have $\tan\theta = x + \tan\alpha$

i.e., $\frac{1}{i} \cdot \frac{e^{i\theta} - e^{-i\theta}}{e^{i\theta} + e^{-i\theta}} = x + \frac{\sin\alpha}{\cos} = \frac{x\cos\alpha + \sin\alpha}{\cos\alpha}$

or $\frac{e^{i\theta} - e^{-i\theta}}{e^{i\theta} - e^{-i\theta}} = \frac{ix\cos\alpha + i\sin\alpha}{\cos\alpha}$

Applying componendo and dividendo, we have

$$\frac{2e^{i\theta}}{2e^{-i\theta}} = \frac{\cos\alpha + (ix\cos\alpha + i\sin\alpha)}{\cos\alpha - (ix\cos\alpha + i\sin\alpha)} = \frac{ix\cos\alpha + (\cos\alpha + i\sin\alpha)}{(\cos\alpha - i\sin\alpha) - ix\cos\alpha}$$

or $e^{2i\theta} = \frac{ix\cos\alpha + e^{i\alpha}}{e^{-i\alpha} - ix\cos\alpha} = \frac{e^{i\alpha}\left(1 + ixe^{-i\alpha}\cos\alpha\right)}{e^{-i\alpha}\left(1 - ixe^{i\alpha}\cos\alpha\right)}$.

Taking logarithm of both sides, we get

$2i\theta = 2i\alpha + \log(1 + ix\, e^{-i\alpha}\cos\alpha) - \log(1 - ix\, e^{i\alpha}\cos\alpha)$

$$= 2i\alpha + \left[ixe^{-i\alpha}\cos\alpha - \frac{i^2x^2e^{-2i\alpha}\cos^2\alpha}{2} + \frac{i^3x^3e^{-3i\alpha}\cos^3\alpha}{3} - \ldots\right]$$

$$-\left[-ixe^{i\alpha}\cos\alpha - \frac{i^2x^2e^{2i\alpha}\cos^2\alpha}{2} - \frac{i^3x^3e^{3i\alpha}\cos^3\alpha}{2} - \ldots\right]$$

$$= 2i\alpha + ix\cos\alpha.\left(e^{i\alpha} + e^{-i\alpha}\right) + \frac{1}{2}i^2x^2\cos^2\alpha\left(e^{2i\alpha} - e^{-2i\alpha}\right)$$

$$+ \frac{1}{3}i^3x^3\cos^3\alpha\left(e^{3i\alpha} - e^{-3i\alpha}\right) + \ldots$$

$$= 2i\alpha + ix\cos\alpha\,(2\cos\alpha) - \frac{1}{2}x^2\cos^2\alpha\,(2i\sin 2\alpha) - \frac{1}{3}ix^3$$

$$\cos^3\alpha\,(2\cos 3\alpha) + \ldots$$

or $2i\theta = 2i\,[\alpha + x\cos^2\alpha - \frac{1}{2}x^2\cos^2\alpha\sin 2\alpha - \frac{1}{3}x^3\cos^3\alpha$

$\cos 3\alpha + \ldots]$

or $\theta = \alpha + \cos^2\alpha - \frac{1}{2}x^2\cos^2\sin 2\alpha - \frac{1}{3}x^3\cos^3\alpha\cos 3\alpha + \ldots \infty.$

Example 16:

If $\cot\phi = (1/x) + \cot\theta$, *prove that*

$$\phi = \frac{x}{\sin\theta}\sin\theta - \frac{1}{2}\frac{x^2}{\sin^2\theta}.\sin 2\theta + \frac{1}{3}\frac{x^3}{\sin^3\theta}\sin 3\theta - \ldots$$

Solution:

We have

$$\cot\phi = \frac{1}{x} + \cot\theta \text{ or } \frac{\cos\phi}{\sin\phi} = \frac{1}{x} + \frac{\cos\theta}{\sin\theta}$$

$$\text{or } \frac{\left(e^{i\phi} + e^{-i\phi}\right)/2}{\left(e^{i\phi} - e^{-i\phi}\right)/2i} = \frac{\sin\theta + x\cos\theta}{x\sin\theta}$$

$$\text{or } \frac{e^{i\phi} + e^{-i\phi}}{e^{i\phi} - e^{-i\phi}} = \frac{\sin\theta + x\cos\theta}{ix\sin\theta}.$$

Applying componedo and dividendo, we get

$$\frac{\left(e^{i\phi} + e^{-i\phi}\right) + \left(e^{i\phi} - e^{-i\phi}\right)}{\left(e^{i\phi} + e^{-i\phi}\right) - \left(e^{i\phi} - e^{-i\phi}\right)} = \frac{(\sin\theta + x\cos\theta) + ix\sin\theta}{(\sin\theta + x\cos\theta) - ix\sin\theta}$$

$$\text{or } \frac{2e^{i\phi}}{2e^{-i\phi}} = \frac{\sin\theta + x(\cos\theta + i\sin\theta)}{\sin\theta + x(\cos\theta - i\sin\theta)} = \frac{\sin\theta + xe^{i\theta}}{\sin\theta + xe^{-i\theta}}$$

$$\text{or } e^{2i\phi} = \frac{\sin\theta\left[1 + \left(\frac{x}{\sin\theta}\right)e^{i\theta}\right]}{\sin\theta\left[1 + \left(\frac{x}{\sin\theta}\right)e^{-i\theta}\right]} = \frac{1 + Ae^{i\theta}}{1 + Ae^{-i\theta}}, \text{ where } A = \frac{x}{\sin\theta}.$$

Taking logarithm of both sides, we have

$2i\phi = \log(1 + Ae^{i\theta}) - \log(1 + Ae^{-i\theta})$

$$= \left(Ae^{i\theta} - \frac{1}{2}A^2e^{2i\theta} + \frac{1}{3}A^3e^{3i\theta} - \ldots\right) - \left(Ae^{-i\theta} - \frac{1}{2}A^2e^{-2i\theta} + \frac{1}{3}A^3e^{-3i\theta} - \ldots\right)$$

$$= A\left(e^{i\theta} - e^{-i\theta}\right) - \frac{1}{2}A^2\left(e^{2i\theta} - e^{-2i\theta}\right) + \frac{1}{3}A^3\left(e^{3i\theta} - e^{-3i\theta}\right) - \ldots$$

$$= A(2i\sin\theta) - \frac{1}{2}A^2(2i\sin 2\theta) + \frac{1}{3}A^3(2i\sin 3\theta) - \ldots$$

$$\text{or } \phi = A\sin\theta - \frac{1}{2}A^2\sin 2\theta + \frac{1}{3}A^3\sin 3\theta - \ldots$$

$$\phi = \frac{x}{\sin\theta}\cdot\sin\theta - \frac{1}{2}\frac{x^2}{\sin^2\theta}\sin 2\theta + \frac{1}{3}\frac{x^3}{\sin^3\theta}\sin 3\theta - \ldots$$

[Putting $A = (x/\sin\theta)$]

EXPANSIONS

4. *Expansion of* $\dfrac{x \sin\theta}{1-2x\cos\theta+x^2}$ *in a series of multiples of* θ *where* $x < 1$.

or Expansion of $\dfrac{x \sin\theta}{1-2x\cos\theta+x^2}$ *in a series of ascending powers of* x.

We know that

$\sin\theta = (e^{i\theta} - e^{-i\theta})/2i$ and $\cos\theta = (e^{i\theta} + e^{-i\theta})/2$.

$$\therefore \frac{x\sin\theta}{1-2x\cos\theta+x^2} = \frac{x\{(e^{i\theta}-e^{-i\theta})/2i\}}{1-2x\cdot\frac{1}{2}(e^{i\theta}+e^{-i\theta})+x^2}$$

$$= \frac{1}{2i}\cdot\frac{x(e^{i\theta}-e^{-i\theta})}{1-x(e^{i\theta}+e^{-i\theta})+x^2} = \frac{1}{2i}\cdot\frac{x(e^{i\theta}-e^{-i\theta})}{(1-xe^{i\theta})(1-xe^{-i\theta})} \qquad \textbf{(Note)}$$

$$= \frac{1}{2i}\left[\frac{1}{(1-xe^{i\theta})}-\frac{1}{(1-xe^{-i\theta})}\right], \qquad \text{(by partial fractions)}$$

$$= \left(\frac{1}{2i}\right)[(1-x\,e^{i\theta})^{-1} - (1-x\,e^{-i\theta})^{-1}]$$

$$= \left(\frac{1}{2i}\right)[(1+x\,e^{i\theta}+x^2\,e^{2i\theta}+x^3\,e^{3i\theta}+\ldots)$$
$$-(1+xe^{-i\theta}+x^2\,e^{-2i\theta}+x^3\,e^{-3i\theta}+\ldots)],$$

[expanding both $(1-xe^{i\theta})^{-1}$ and $(1-xe^{-i\theta})^{-1}$ by binomial theorem which is justified because $|xe^{i\theta}| = |x|\,|e^{i\theta}| = |x|\cdot|\cos\theta + i\sin\theta| = |x| < 1$, x being given to be < 1, and similarly $|xe^{-i\theta}| < 1$]

$$= \frac{1}{2i}\left[x(e^{i\theta}-e^{-i\theta})+x^2(e^{2i\theta}-e^{-2i\theta})+x^3(e^{3i\theta}-e^{-3i\theta})+\ldots\right]$$

$$= x\left(\frac{e^{i\theta}-e^{-i\theta}}{2i}\right)+x^2\left(\frac{e^{2i\theta}-e^{-2i\theta}}{2i}\right)+x^3\left(\frac{e^{3i\theta}-e^{-3i\theta}}{2i}\right)+\ldots$$

$= x\sin\theta + x^2\sin 2\theta + x^3\sin 3\theta + \ldots$ ad.inf.

5. *If* $x < 1$, *prove that*

$$\frac{x\cos\theta}{1-2x\sin\theta+x^2} = x\cos\theta + x^2\sin 2\theta - x^3\cos 3\theta - x^4\sin 4\theta + \ldots \textit{ ad.inf.}$$

We have

$$\frac{x\cos\theta}{1-2x\sin\theta+x^2}=\frac{x.\frac{1}{2}\left(e^{i\theta}+e^{-i\theta}\right)}{1-2x\left\{\left(e^{i\theta}-e^{-i\theta}\right)/2i\right\}+x^2}$$

$$=\frac{1}{2}\frac{x\left(e^{i\theta}+e^{-i\theta}\right)}{1+ix\left(e^{i\theta}-e^{-i\theta}\right)+x^2} \qquad \left[\because \frac{1}{i}=-i\right]$$

$$=\frac{1}{2}\frac{x\left(e^{i\theta}+e^{-i\theta}\right)}{1+ix\left(e^{i\theta}-e^{-i\theta}\right)-i^2x^2e^{i\theta}e^{-i\theta}} \qquad \textbf{(Note)}$$

$$=\frac{1}{2}\frac{x\left(e^{i\theta}+e^{-i\theta}\right)}{1+ix\left(e^{i\theta}-e^{-i\theta}\right)-ixe^{i\theta}.ixe^{-i\theta}}=\frac{1}{2i}\frac{ix\left(e^{i\theta}+e^{-i\theta}\right)}{\left(1+ixe^{i\theta}\right)\left(1-ixe^{-i\theta}\right)}$$

$$=\frac{1}{2i}\left[\frac{1}{1-ixe^{-i\theta}}-\frac{1}{1+ixe^{i\theta}}\right], \qquad \text{(by partial fractions)}$$

$$=\frac{1}{2i}\left[\left(1-ixe^{-i\theta}\right)^{-1}-\left(1+ixe^{i\theta}\right)^{-1}\right].$$

The binomial expansions of

$(1 - ix\ e^{-i\theta})^{-1}$ and $(1 + ix\ e^{i\theta})^{-1}$

are valid since $|ix\ e^{-i\theta}| = |x| < 1$

and similarly $|ix\ e^{i\theta}| = |x| < 1$, x being given to be < 1.

Hence $\dfrac{x\cos\theta}{1-2x\sin\theta+x^2}$

$$=\left(\frac{1}{2i}\right)[(1 + ix\ e^{-i\theta} + i^2x^2\ e^{-2i\theta} + i^3x^3e^{-3i\theta} + \ldots) - (1 - ix\ e^{i\theta} + i^2x^2\ e^{2i\theta} - i^3x^3e^{3i\theta} + \ldots)]$$

$$=\left(\frac{1}{2i}\right)[(ix\ (e^{i\theta} + e^{-i\theta}) - i^2x^2\ (e^{2i\theta} - e^{-2i\theta}) + i^3x^3\ (e^{3i\theta} + e^{-3i\theta}) - i^4x^4\ (e^{4i\theta} - e^{-4i\theta}) + \ldots]$$

$$=x\left(\frac{e^{i\theta}+e^{-i\theta}}{2}\right)+x^2\left(\frac{e^{2i\theta}-e^{-2i\theta}}{2i}\right)-x^3\left(\frac{e^{3i\theta}+e^{-3i\theta}}{2}\right)-x^4\left(\frac{e^{4i\theta}-e^{-4i\theta}}{2i}\right)+\ldots$$

$= x \cos\theta + x^2 \sin 2\theta - x^3 \cos 3\theta - x^4 \sin 4\theta + \ldots$

6. *Expansion of* $\frac{\sin n\theta}{\sin\theta}$ *in a series of descending powers of* $\cos\theta$.

in 4 we have proved that

$$\frac{x\sin\theta}{1-2x\cos\theta+x^2} = x\sin\theta + x^2\sin 2\theta + \ldots + x^n\sin n\theta + \ldots$$

Dividing both sides by x sin θ, we have

$$\frac{1}{1-2x\cos\theta+x^2} = 1+\frac{x\sin 2\theta}{\sin\theta}+\frac{x^2\sin 3\theta}{\sin\theta}+\ldots+\frac{x^{n-1}\sin n\theta}{\sin\theta}+\ldots \qquad \ldots(1)$$

We observe that $\frac{\sin n\theta}{\sin\theta}$ is the coefficient of x^{n-1} in the R.H.S. of (1).

$\therefore \frac{\sin n\theta}{\sin\theta}$ = the coefficient of x^{n-1} in the R.H.S. of (1)

= the coeff. of x^{n-1} in the expansion of $\frac{1}{1-2x\cos\theta+x^2}$

= the coeff. of x^{n-1} in the expansion of $[1 - 2x\cos\theta + x^2]^{-1}$

= the coeff. of x^{n-1} in the expansion of $[1 - x(2\cos\theta - x)]^{-1}$

= the coeff. of x^{n-1} in $[1 + x(2\cos\theta - x) + x^2(2\cos\theta - x)^2$

$+ x^3(2\cos\theta - x)^3 + \ldots + x^{n-1}(2\cos\theta - x)^{n-1} + \ldots\infty]$.

Now the terms after $x^{n-1}(2\cos\theta - x)^{n-1}$ contain x^n and higher powers of x which we do not require.

The coeff. of x^{n-1}

in $x^{n-1}(2\cos\theta - x)^{n-1}$ is $(2\cos\theta)^{n-1}$,

in $x^{n-2}(2\cos\theta - x)^{n-2}$ is $-(n-2)(2\cos\theta)^{n-3}$,

in $x^{n-3}(2\cos\theta - x)^{n-3}$ is $\frac{(n-3)(n-4)}{1.2}(2\cos\theta)^{n-5}$,

and so on.

Hence $\frac{\sin n\theta}{\sin\theta} = (2\cos\theta)^{n-1} - (n-2)(2\cos\theta)^{n-3} + \frac{(n-3)(n-4)}{1.2}(2\cos\theta)^{n-5} - \ldots$,

the last term being $(-1)^{(n-2)/2}(2\cos\theta)$ if n is even

and $(-1)^{(n-1)/2}$ if n is odd.

7. *Expansion of* $\frac{1-x^2}{1-2x\cos\theta+x^2}$ *in a series of cosines of multiples of* θ, $x < 1$.

Or

Expand $\frac{1-x^2}{1-2x\cos\theta+x^2}$ *in a series of ascneding powers of x.*

Here the highest power of x in the numerator and denominator is the same (*i.e.*, 2), therefore we shall divide the Nr. by the Dr. to make the highest power of x in the Nr. les than that in the Dr.

$$\therefore \frac{1-x^2}{1-2x\cos\theta+x^2} = -1 + \frac{2-2x\cos\theta}{1-2x\cos\theta+x^2}$$

$$= -1 + \frac{2-2x.\frac{1}{2}\left(e^{i\theta}+e^{-i\theta}\right)}{1-2x.\frac{1}{2}\left(e^{i\theta}+e^{-i\theta}\right)+x^2} = -1 + \frac{2-x\left(e^{i\theta}+e^{-i\theta}\right)}{\left(1-xe^{i\theta}\right)\left(1-xe^{-i\theta}\right)}$$

$$= -1 + \frac{\left(1-xe^{i\theta}\right)+\left(1-xe^{-i\theta}\right)}{\left(1-xe^{i\theta}\right)\left(1-xe^{-i\theta}\right)} = -1 + \frac{1}{\left(1-xe^{-i\theta}\right)} + \frac{1}{\left(1-xe^{i\theta}\right)}$$

$= -1 + (1 - xe^{-i\theta})^{-1} + (1 - xe^{-i\theta})^{-1}$

$= -1 + (1 + xe^{-i\theta} + x^2e^{-2i\theta} + \ldots) + (1 + xe^{i\theta} + x^2e^{2i\theta} + \ldots)$,

expanding the terms on the R.H.S. by binomial expansion, x being < 1

$= 1 + x\,(e^{i\theta} + e^{-i\theta}) + x^2\,(e^{2i\theta} + e^{-2i\theta}) + \ldots$

$= 1 + 2x\cos\theta + 2x^2\cos 2\theta + \ldots + 2x^n\cos n\theta + \ldots$ ad.inf.

8. *Expansion of cos* $n\theta$ *in a series of descending powers of cos* θ

In the previous article (*i.e.*, in 7), we have proved that

$$\frac{1-x^2}{1-2x\cos\theta+x^2} = 1 + 2x\cos\theta + 2x^2\cos 2\theta + \ldots + 2x^n\cos n\theta + \ldots$$

$\therefore$ $2\cos n\theta$ = the coefficient of x^n in the R.H.S.

= the coefficient of x^n in $\frac{1-x^2}{1-2x\cos\theta+x^2}$

= the coeffi. of x^n in $(1 - 2x\cos\theta + x^2)^{-1}$

– the coeff. of x^{n-2} in $(1 - 2x\cos\theta + x^2)^{-1}$.

Now $(1 - 2x\cos\theta + x^2)^{-1} = [1 - x\,(2\cos\theta - x)]^{-1}$

$= 1 + x(2\cos\theta - x) + x^2(2\cos\theta - x)^2 + \ldots$

$+ x^n(2\cos\theta - x)^n + \ldots \qquad \ldots(1)$

From the R.H.S. of (1), the coeff. of x^n

in $x^n(2\cos\theta - x)^n$ is $(2\cos\theta)^n$,

in $x^{n-1}(2\cos\theta - x)^{n-1}$ is $-(n-1)(2\cos\theta)^{n-2}$,

in $x^{n-2}(2\cos\theta - x)^{n-2}$ is $\dfrac{(n-2)(n-3)}{2!}(2\cos\theta)^{n-4}$,

in $x^{n-3}(2\cos\theta - x)^{n-3}$ is $\dfrac{-(n-3)(n-4)(n-5)}{3!}(3\cos\theta)^{n-6}$,

and so on.

Adding all these we shall get the coefficient of x^n in

$(1 - 2x\cos\theta + x^2)^{-1}$.

Again from the R.H.S. of (1), the coeff. of x^{n-2}

in $x^{n-2}(2\cos\theta - x)^{n-2}$ is $(2\cos\theta)^{n-2}$,

in $x^{n-3}(2\cos\theta - x)^{n-3}$ is $-(n-3)(2\cos\theta)^{n-4}$,

in $x^{n-4}(2\cos\theta - x)^{n-4}$ is $\dfrac{(n-4)(n-5)}{2!}(2\cos\theta)^{n-6}$,

and so on.

Adding all these we shall get the coeff. of

x^{n-2} in $(1 - 2x\cos\theta + x^2)^{-1}$.

Hence $2\cos n\theta = [(2\cos\theta)^n - (n-1)(2\cos\theta)^{n-2}$

$+ \{(n-2)(n-3)/2!\}(2\cos\theta)^{n-4}$

$- \{(n-3)(n-4)(n-5)/3!\}(2\cos\theta)^{n-6} + \ldots]$

$- [(2\cos\theta)^{n-2} - (n-3)(2\cos\theta)^{n-4} + \{(n-4)(n-5)/2!\}$

$(2\cos\theta)^{n-6} - \ldots]$

$$= (2\cos\theta)^n - n(2\cos\theta)^{n-2} + \left[\frac{(n-2)(n-3)}{2!} + (n-3)\right](2\cos\theta)^{n-4}$$

$$-\left[\frac{(n-3)(n-4)(n-5)}{3!} + \frac{(n-4)(n-5)}{2!}\right](2\cos\theta)^{n-6} + \ldots$$

$$= (2\cos\theta)^n - n(2\cos\theta)^{n-2} + \frac{n(n-3)}{2!}(2\cos\theta)^{n-4} - \frac{n(n-4)(n-5)}{3!}(2\cos\theta)^{n-6} + \ldots$$

The last term being $(-1)^{(n-1)/2}$ $(2\cos\theta)$ n, if n is odd and $(-1)^{n/2}.2$, if n is even.

9. *Expansion of log $(1 - 2x\cos\theta + x^2)$ in a series of cosines of multiples of θ.*

We have $\log(1 - 2x\cos\theta + x^2)$

$$= \log[1 - x(e^{i\theta} + e^{-i\theta}) + x^2] \qquad \left[\because \cos\theta = \frac{1}{2}\left(e^{i\theta} + e^{-i\theta}\right)\right]$$

$$= \log[(1 - xe^{i\theta})(1 - xe^{-i\theta})] = \log(1 - xe^{i\theta}) + \log(1 - xe^{-i\theta})$$

$$= \left(-xe^{i\theta} - \frac{1}{2}x^2e^{2i\theta} - \frac{1}{3}x^3e^{3i\theta} - \ldots\right) + \left(-xe^{-i\theta} - \frac{1}{2}x^2e^{-2i\theta} - \frac{1}{3}x^3e^{-3i\theta} - \ldots\right),$$

expanding both $\log(1 - xe^{i\theta})$ and $\log(1 - xe^{-i\theta})$ by logarithmic expansion which is justified if $|x| > 1$

$$= -x(e^{i\theta} + e^{-i\theta}) - \frac{1}{2}x^2\left(e^{2i\theta} + e^{-2i\theta}\right) - \frac{1}{3}x^3\left(e^{3i\theta} + e^{-3i\theta}\right) - \ldots$$

$$= -2\left[x\cos\theta + \frac{1}{2}x^2\cos 2\theta + \frac{1}{3}x^3\cos 3\theta + \ldots\text{ad.inf.}\right]$$

10. *If $\sin x = n\sin(a + x)$, expand x in a series of ascending powers of n, where n is less than unity.*

We have $\sin x = n\sin(a + x)$.

$$\therefore \frac{e^{ix} - e^{-ix}}{2i} = n.\frac{e^{i(a+x)} - e^{-i(a+x)}}{2i} \qquad \left(\because \sin\theta = \frac{e^{i\theta} - e^{-i\theta}}{2i}\right)$$

or $(e^{ix} - e^{-ix}) = n[e^{i(a+x)} - e^{-i(a+x)}]$

or $e^{2ix} - 1 = n(e^{ia}.e^{2ix} - e^{-ia})$ [Multiplying both sides by e^{ix}]

or $e^{2ix}(1 - ne^{ia}) = (1 - ne^{-ia})$

or $e^{2ix} = \{(1 - ne^{-na})/(1 - ne^{ia})\}$.

Taking logarithm of both sides, we have

$2ix = \log(1 - ne^{-ia}) - \log(1 - ne^{ia})$

$$= \left(-ne^{-ia} - \frac{1}{2}n^2e^{-2ia} - \frac{1}{3}n^3e^{-3ia} - \ldots\right)$$

$$-\left(-ne^{ia} - \frac{1}{2}n^2e^{2ia} - \frac{1}{3}n^3e^{3ia} - \ldots\right),$$

the logarithmic expansions are valid because $|ne^{-ia}| = |n| < 1$ and also $|ne^{ia}| = |n| < 1$.

$$= n\left(e^{ia} - e^{-ia}\right) + \frac{1}{2}n^2\left(e^{2ia} - e^{-2ia}\right) + \frac{1}{3}n^3\left(e^{3ia} - e^{-3ia}\right) + \ldots$$

$$\therefore\ x = n\left(\frac{e^{ia} - e^{-ia}}{2i}\right) + \frac{1}{2}n^2\left(\frac{e^{2ia} - e^{-2ia}}{2i}\right) + \frac{1}{3}n^3\left(\frac{e^{3ia} - e^{-3ia}}{2i}\right) + \ldots$$

$$= n \sin a + \frac{1}{2}n^2 \sin 2a + \frac{1}{3}n^3 \sin 3a + \ldots$$

EXERCISES

1. Prove that $\frac{\pi}{4} = \left[\frac{2}{3} + \frac{1}{7}\right] - \frac{1}{3}\left[\frac{2}{3^3} + \frac{1}{7^3}\right] + \frac{1}{5}\left[\frac{2}{3^5} + \frac{1}{7^5}\right] - \ldots$

2. Prove that $\frac{\pi}{8} = \frac{1}{1.3} + \frac{1}{5.7} + \frac{1}{9.11} + \ldots \text{ad.inf.}$ **(Garhwal, 1991)**

3. Show that $\frac{\pi}{12} = \left\{1 - \frac{1}{3^{1/2}}\right\} - \frac{1}{2}\left\{1 - \frac{1}{3^{3/2}}\right\} + \frac{1}{5}\left\{1 - \frac{1}{3^{5/2}}\right\} - \ldots\infty.$

4. Show that $\frac{\pi}{4} = \left[\frac{1}{2} + \frac{1}{5} + \frac{1}{8}\right] - \frac{1}{3}\left[\frac{1}{2^3} + \frac{1}{5^3} + \frac{1}{8^3}\right] + \frac{1}{5}\left[\frac{1}{2^5} + \frac{1}{5^5} + \frac{1}{8^5}\right] - \ldots$

5. If $x < \left(\sqrt{2} - 1\right)$, prove that $2\left(x - \frac{1}{3}x^3 + \frac{1}{5}x^5 - \ldots\text{ad.inf.}\right)$

$$= \left(\frac{2x}{1 - x^2}\right) - \frac{1}{3}\left(\frac{2x}{1 - x^2}\right)^3 + \frac{1}{5}\left(\frac{2x}{1 - x^2}\right)^5 - \ldots$$ **(Bundelkhnd, 1990)**

6. If $x > 0$, prove that $\tan^{-1} x = \frac{\pi}{4} + \frac{x - 1}{x + 1} - \frac{1}{3}\left\{\frac{x - 1}{x + 1}\right\}^3 + \ldots$ **(Agra, 1996)**

7. When θ lies between 0 anad $\frac{1}{2}\pi$, prove that

$$\tan^{-1}\left(\frac{1-\cos\theta}{1+\cos\theta}\right) = \tan^2\frac{\theta}{2} - \frac{1}{3}\tan^6\frac{\theta}{2} + \frac{1}{5}\tan^{10}\frac{\theta}{2} - \ldots\infty.$$

8. When both θ and $\tan^{-1}$ (sec θ) lie between 0 and $\frac{1}{2}\pi$, prove that

$$\tan^{-1}(\sec\theta) = \frac{1}{4}\pi + \tan^2\frac{1}{2}\theta - \frac{1}{3}\tan^6\frac{1}{2}\theta + \frac{1}{5}\tan^{10}\frac{1}{2}\theta - \ldots$$

9. Prove that $\tan^{-1}\left(\frac{\cos\theta+\sin\theta}{\cos\theta-\sin\theta}\right) = n\pi + \frac{1}{4}\pi + \tan\theta - \left(\frac{1}{3}\right)\tan^3\theta + \left(\frac{1}{5}\right)$ $\tan^5\theta - \ldots$

10. If x lies between $-\frac{\pi}{4}$ and $\frac{\pi}{4}$, show that $\tan x - \frac{1}{3}\tan^3 x + \frac{1}{5}\tan^5 x - \ldots \text{ad.inf.}$

$$= \tanh x + \frac{1}{3}\tanh^3 + \frac{1}{5}\tanh^5 x + \ldots \text{ad.inf.}$$ **(Meerut, 1986)**

11. If $\tan x < 1$, show that $\tan^2 x - \frac{1}{2}\tan^4 x + \frac{1}{3}\tan^6 x - \ldots = \sin^2 x + \frac{1}{2}\sin^4 x + \frac{1}{3}\sin^6 x + \ldots$

12. Prove that if x, y, z are cube roots of unity,

$$\frac{\tan^{-1}x}{x} + \frac{\tan^{-1}y}{y} + \frac{\tan^{-1}z}{z} = 3\left[1 - \frac{1}{7} + \frac{1}{13} - \frac{1}{19} + \frac{1}{25} - \ldots\right].$$

7

Summation of Trigonometrical Series

7.1 INTRODUCTION

In this method we shall deal with the summation of some trigonometrical series with a finite and infinite number of terms. It may be noted that all the series with infinite number of terms are not summable and hence due care should be taken regarding the convergence of such series.

7.2 GENERAL C + IS METHOD

A general method for the Summation of Trigonometrical Series of the type:

$$C_1 \cos \theta + C_2 \cos (\theta + \alpha) + C_3 \cos (\theta + 2\alpha) +$$

Or $$C_1 \sin \theta + C_2 \sin (\theta + \alpha) + C_3 \sin (\theta + 2\alpha) +$$

is to denote the serie (1) by C, since the coefficients of C's in this series are cosines of the angles increasing in arithmetic progression and the series (2) by S being a sine series. Multiplying (2) by i and adding

$$C + is = C_1 (\cos \theta + i \sin \theta) + C_2 \{\cos (\theta + \alpha) + i \sin (\theta + \alpha)\}$$
$$+ C_3 \cos (\theta + 2\alpha) + i \sin (\theta + 2\alpha) +$$

$$= C_1 e^i \theta + C_2 e^i (\theta + \alpha) C_3 e^i (\theta + 2\alpha) +$$

$$= e^i \theta \{C_1 + C_1 e^{i\alpha} + C_3 e^{2i\alpha} + ...\}$$

Now the series within the bracket of (3) may be either:

(i) Series in Geometrical progression.

(ii) Binomial series or one which can be reduced to it.

(iii) Exponential series or the allied series *e.g.*, sine or cosine series, and

(iv) Logarithmic series or the allied series *e.g.*, Gregory's series.

The real part of (3) gives C and the imaginary part gives S. If either of the C or S is given, the other series known as auxilliary series can be formed and the sum of C+ is found the real part of the sum. So found is equal to C and the imaginary part equal to so.

7.3 USE OF GEOMATRIC SERIES

The geomatric series is given by

$$a + az + az^2 + az^3 + ... a_n \text{ inf}$$

sums of n terms and infinity of the above series are given by

$$S_n = \frac{a(z^n - 1)}{z - 1} \text{ if } |z| > 1$$

$$= \frac{a(1 - z^n)}{1 - z} \text{ if } |z| < 1$$

$$S_\infty = \frac{a}{1 - z} \text{ provided } |z| < 1.$$

7.4 SUMMATION OF SERIES

The following formulae are very useful in solving problems of Summation of trigonometrical series:

1. $e^{ix} = \cos x + i \sin x.$ **[Euler's formula]**
2. $e^{-ix} = \cos x - i \sin x.$
3. $e^{ix} + e^{-ix} = 2 \cos x,\ e^{ix} - e^{-ix} = 2i \sin x.$
4. $e^{inx} = \cos nx + i \sin nx$, n is an integer.
5. $(1 + a)^n = 1 + na + \frac{n(n-1)}{2!} a^2 + \frac{n(n-1)(n-2)}{3!} a^3 + ... + a^n$
6. $e^x = 1 + x + \frac{x^2}{2!} + \frac{x^3}{3!} + ...$
7. $1 + x + x^2 + ... + x^{n-1} = \frac{1 - x^n}{1 - x},\ x \neq 1.$

The following rule is used in the Summation of Trigonometrical Series:

Rule: *Write the cosines series as C and the sine series as S. Simplify C + iS use the above expansions.*

Example 1:

Find the sum of series:

$$\cos\theta - \frac{1}{2}\cos 2\theta + \frac{1}{3}\cos 3\theta - \ldots \text{ ad inf.}$$

$$\sin\theta - \frac{1}{2}\sin 2\theta + \frac{1}{3}\sin 3\theta - \ldots \text{ ad inf.}$$

Solution:

Let $C = \cos\theta - \frac{1}{2}\cos 2\theta + \frac{1}{3}\cos 3\theta - \ldots..$

and $S = \sin\theta - \frac{1}{2}\sin 2\theta + \frac{1}{3}\sin 3\theta - \ldots.$

$\therefore\ C + iS = (\cos\theta + i\sin\theta) - \frac{1}{2}(\cos 2\theta + i\sin 2\theta) + \ldots.$

$= (\cos\theta + i\sin\theta) - \frac{1}{2}(\cos\theta + i\sin\theta)^2 + \frac{1}{3}(\cos\theta + i\sin\theta)^3 - \ldots..$

$= z - \frac{1}{2}z^2 + \frac{1}{3}z^3 - \ldots.,\ z = \cos\theta + i\sin\theta$

$= \log(1 + z)$

$= \log(1 + \cos\theta + i\sin\theta)$

$= \log(2\cos^2\theta/2 + 2i\sin\theta/2\cos\theta/2)$

$= \log[2\cos\theta/2(\cos\theta/2 + i\sin\theta/2)]$

$= \log[2\cos\theta/2.e^{i\theta/2}]$

$= \log(2\cos\theta/2) + \log(e^{i\theta/2})$

$= \log(2\cos\theta/2) + i\theta/2.$

Comparing real and imaginary parts on both sides, we get

$C = \log(2\cos\theta/2),$

$S = \frac{1}{2}\theta.$

Example 2:

Sum the series:

$$\sin\alpha + c\sin(\alpha + \beta) + \frac{c^2}{2!}\sin(\alpha + 2\beta) + \ldots + \text{ad. inf.}$$

Solution:

Let $S = \sin\alpha + c\sin(\alpha + \beta) + \frac{c^2}{2!}\sin(\alpha + 2\beta) + \ldots\infty$

and $C = \cos\alpha + c\cos(\alpha + \beta) + \frac{c^2}{2!}\cos(\alpha + 2\beta) + \ldots\infty$

$$\therefore C + iS = (\cos\alpha + i\sin\alpha) + c\{\cos(\alpha+\beta) + i\sin(\alpha+\beta)\}$$
$$+ \frac{c^2}{2!}\{\cos(\alpha+2\beta) + i\sin(\alpha+2\beta)\} + \dots\infty$$

$$= e^{i\alpha} + c\,e^{i(\alpha+\beta)} + \frac{c^2}{2!}e^{i(\alpha+2\beta)} + \dots\infty$$

$$= e^{i\alpha}\left[i + c\,e^{i\beta} + \frac{c^2}{2!}e^{2i\beta} + \dots\infty\right]$$

$$= e^{i\alpha}\left[i + cz + \frac{1}{2\,i}c^2z^2 + \frac{1}{3\,i}c^3z^3 + \dots\infty\right],\ z = e^{i\beta}$$

$$= e^{i\alpha}e^{cz} = (\cos\alpha + i\sin\alpha)\,e^{c(\cos\beta + i\sin\beta)}$$

$$= e^{c\cos\beta}(\cos\alpha + i\sin\alpha)\,e^{ic\sin\beta}$$

$$C + iS = e^{c\cos\beta}(\cos\alpha + i\sin\alpha)\{\cos(c\sin\beta) + i\sin(c\sin\beta)\}$$

Equating imaginary parts on both sides, we obtain

$$S = e^{c\cos\beta}\{\sin\alpha\cos(c\sin\beta) + \cos\alpha\sin(c\sin\beta)\}$$

Hence $S = e^{c\cos\beta}\sin(\alpha + c\sin\beta)$.

Example 3:

Prove that if n is a positive integer, then

(i) $\cos\theta + {}^nc_1\cos(\theta+\phi) + {}^nc_1\cos(\theta+2\phi) + \dots + {}^nc_n\cos(\theta+n\phi)$

$$= (2\cos\phi/2)^n\cos\left(\theta + \frac{1}{2}n\phi\right).$$

(ii) $\sin\theta + {}^nc_1\sin(\theta+\phi) + {}^nc_2\sin(\theta+2\phi) + \dots + {}^nc_n\sin(\theta+n\phi)$

$$= (2\cos\phi/2)^n\sin\left(\theta + \frac{1}{2}n\phi\right).$$

Solution:

Let $C = \cos\theta + {}^nc_1\cos(\theta+\phi) + {}^nc_2\cos(\theta+2\phi)$
$+ \dots + {}^nc_n\sin(\theta+n\phi)$.

$$\therefore C + iS = e^{i\theta} + {}^nc_1 e^{i(\theta+\phi)} + {}^nc_2 e^{i(\theta-2\phi)} + \dots + {}^nc_n e^{i(\theta+n\phi)}.$$
$$= e^{i\theta}[1 + {}^nc_1 e^{i\phi} + {}^nc_2 e^{i2\phi} + \dots + {}^nc_n e^{in\phi}]$$
$$= e^{i\theta}(1 + e^{i\phi})^n$$
$$= e^{i\theta}[1 + \cos\phi + i\sin\phi]^n$$

$= e^{i\theta}\,[2\cos^2\phi/2 + 2i\sin\phi/2\cos\phi/2]^n$

$e^{i\theta}\,(2\cos\phi/2)^n\,[\cos\phi/2 + i\sin\phi/2]^n$

$\therefore C + iS = (2\cos\phi/2)^n(\cos\theta + i\sin\theta)(\cos n\phi/2 + i\sin n\phi/2)$ ...(1)

Comparing real parts on both sides,we get

$$C = (2\cos\phi/2)^n\,[\cos\theta\cos n\phi/2 - \sin\theta\sin n\phi/2]$$

Hence $\quad C = (2\cos\phi/2)^n \cos\left(\theta + \frac{1}{2}n\phi\right).$

Similarly on comparing imaginary parts on both sies of (1),

$S = (2\cos\phi/2)^n \sin\left(\theta + \frac{1}{2}n\phi\right).$

Example 4:

Sum to n terms:

$$n\cos\theta + \frac{n(n-1)}{1.2}\cos 2\theta + \frac{n(n-1)(n-2)}{1.2.3}\cos 3\theta + \ldots$$

Solution:

Let

$$C = n\cos\theta + \frac{n(n-1)}{1.2}\cos 2\theta + \frac{n(n-1)(n-2)}{1.2.3}\cos 3\theta + \ldots + \cos n\theta,$$

$$S = n\sin\theta + \frac{n(n-1)}{1.2}\cos 2\theta + \frac{n(n-1)(n-2)}{1.2.3}\sin 3\theta + \ldots + \sin n\theta.$$

$$\therefore C + iS = \left[1 + ne^{i\theta} + \frac{n(n-1)}{1.2}e^{12\theta} + \ldots + e^{in\theta}\right] - 1$$

$= (1 + e^{i\theta})^n - 1$

$= (1 + \cos^2\theta + i\sin\theta)^n - 1$

$= (2\cos^2\theta/2 + 2i\sin\theta/2\cos\theta/2)^n - 1$

$(2\cos\theta/2)^n\,[\cos\theta/2 + i\sin\theta/2]^n - 1$

$= (2\cos\theta/2)^n\,[\cos(n\theta/2) + i\sin(n\theta/2)] - 1.$

Comparing real parts on both sides, we get

$C = (2\cos\theta/2)^n\cos(n\theta/2) - 1.$

Example 5:

Sum to n terms:

(1) cos θ + cos 2θ + cos 3θ + ...

(2) sin θ + sin 2θ + sin 3θ + ...

Solution:

Let $C = \cos\theta + \cos 2\theta + ... + \cos n\theta$,

and $S = \sin\theta + \sin 2\theta + ... + \sin n\theta$.

$\therefore\ C + iS = (\cos\theta + i\sin\theta) + (\cos 2\theta + i\sin 2\theta) + ... + (\cos n\theta + i\sin n\theta)$

$= (\cos\theta + i\sin\theta) + (\cos\theta + i\sin\theta)^2 + ... + (\cos\theta + i\sin\theta)^n$ [by DeMoivre's Theorem]

$= z + z^2 + ... + z^n$, where $z = \cos\theta + i\sin\theta$.

$$= z(1 + z + z^2 + ... + z^{n-1}) = \frac{z(1-z^n)}{1-z}$$

$$= \frac{(\cos\theta + i\sin\theta)\left\{1 - (\cos\theta + i\sin\theta)^n\right\}}{1 - \cos\theta - i\sin\theta}$$

$$= \frac{(\cos\theta + i\sin\theta) - (\cos\theta + i\sin\theta)^{n+1}}{(1-\cos\theta) - i\sin\theta} \times \frac{(1-\cos\theta) + i\sin\theta}{(1-\cos\theta) + i\sin\theta}$$

$$= (\cos\theta + i\sin\theta) - \{\cos(n+1)\theta + i\sin(n+1)\theta\} \times \frac{(1-\cos\theta) + i\sin\theta}{(1-\cos\theta)^2 + i\sin^2\theta}$$

$$= \frac{\left[\{\cos\theta - \cos(n+1)\theta\} + i\{\sin\theta - \sin(n+1)\theta\}\right]\{(1-\cos\theta) + i\sin\theta\}}{2-2\cos\theta}$$

$$= \frac{\begin{array}{c}\left[\{\cos\theta - \cos(n+1)\theta\}(1-\cos\theta) - \{\sin\theta - \sin(n+1)\theta\}\sin\theta\right] \\ +i\left[\sin\theta\{\cos\theta - \cos(n+1)\theta\} + \{\sin\theta - \sin(n+1)\theta\}(1-\cos\theta)\right]\end{array}}{2(1-\cos\theta)}$$

$$= \frac{\begin{array}{c}\left[\cos\theta\{\cos(n+1)\theta\cos\theta + \sin(n+1)\theta\sin\theta\} - \cos(n+1)\theta - 1\right] \\ + i\left[\sin\theta + \{\sin(n+1)\theta\cos\theta - \cos(n+1)\theta\sin\theta\} - \sin(n+1)\theta\right]\end{array}}{2(1-\cos\theta)}$$

Equating real and imaginary parts on both sides, we get

$$C = \frac{\cos\theta + \cos n\theta - \cos(n+1)\theta - 1}{2(1-\cos\theta)}$$

$$S = \frac{\sin\theta + \sin n\theta - \sin(n+1)\theta}{2(1-\cos\theta)}.$$

Example 6:

Sum to n terms

(i) $\cos\alpha + \cos(\alpha+\beta) + \cos(\alpha+2\beta) + \ldots.$

(ii) $\sin\alpha + \sin(\alpha+\beta) + \sin(\alpha+2\beta) + \ldots.,\ \beta \neq 2k\pi.$

Solution:

Let $C = \cos\alpha + \cos(\alpha+\beta) + \ldots. + \cos\{\alpha + (n-1)\beta\}$,

and $S = \sin\alpha + \sin(\alpha+\beta) + \ldots. + \sin\{\alpha + (n-1)\beta\}$.

$\therefore\ C + iS = (\cos\alpha + i\sin\alpha) + \{\cos(\alpha+\beta) + i\sin(\alpha+\beta)\} + \ldots.$
$+ [\cos\{\alpha + (n-1)\beta\} + i\sin\{\alpha + (n-1)\beta\}]$

$= e^{i\alpha} + e^{i(\alpha+\beta)} + e^{i(\alpha+2\beta)} + \ldots.. + e^{i[\alpha+(n-1)\beta]}$

$= e^{i\alpha}\{1 + e^{i\beta} + e^{2i\beta} + \ldots. + e^{(n-1)i\beta}\}$

$= e^{i\alpha}\{1 + z + z^2 + \ldots... + z^{n-1}\},\ z = e^{i\beta} = \cos\beta + i\sin\beta$

$$= e^{ia}\left(\frac{1-z^n}{1-z}\right)$$

$$= \frac{e^{i\alpha}(1-\cos n\beta - i\sin n\beta)}{1-\cos\beta - i\sin\beta}$$

$$= \frac{e^{i\alpha}\{2\sin^2(n\beta/2) - 2i\sin(n\beta/2)\cos(n\beta/2)\}}{2\sin^2(\beta/2) - 2i\sin(\beta/2)\cos(\beta/2)}$$

$$= \frac{-2i\,e^{i\alpha}\sin(n\beta/2)\{\cos(n\beta/2) + i\sin(n\beta/2)\}}{-2i\sin(\beta/2)\{\cos(\beta/2) + i\sin(\beta/2)\}}$$

$$= \frac{\sin(n\beta/2)}{\sin(\beta/2)}(\cos\alpha + i\sin\alpha)\{\cos(n-1)(\beta/2) + i\sin(n-1)(\beta/2)\}$$

$$= \frac{\sin(n\beta/2)}{\sin(\beta/2)}[\cos\{\alpha + (n-1)(\beta/2)\} + i\sin\{\alpha + (n-1)(\beta/2)\}]$$

Equating real and imaginary parts on both sides,

$$C = \frac{\sin(n\beta/2)}{\sin(\beta/2)} \cos\{\alpha + (n-1)(\beta/2)\}$$

$$S = \frac{\sin(n\beta/2)}{\sin(\beta/2)} \sin\{\alpha + (n-1)(\beta/2)\}.$$

Example 7:

Sum to n terms:

$\cos\theta + \cos(\theta + \pi/n) + \cos(\theta + 2\pi/n) + \ldots$

Solution:

Let

$C = \cos\theta + \cos(\theta + \pi/n) + \cos(\theta + 2\pi/n) + \ldots + \cos\{\theta + (n-1)\pi/n\},$

$S = \sin\theta + \sin(\theta + \pi/n) + \sin(\theta + 2\pi/n) + \ldots + \sin\{\theta + (n-1)\pi/n\}.$

$\therefore C + iS = e^{i\theta} + e^{i(\theta + \pi/n)} + e^{i(\theta + 2\pi)} + \ldots$

$= e^{i\theta}[1 + e^{i\pi/n} + e^{i2\pi/n} + \ldots + e^{i(n-1)\pi/n}]$

$$= \frac{e^{i\theta}(1-z^n)}{1-z} = \frac{e^{i\theta}(1-e^{i\pi})}{1-e^{i\pi/n}}$$

$$= \frac{e^{i\theta}[1-(\cos\pi + i\sin\pi)]}{1-(\cos\pi/n + i\sin\pi/n)}$$

$$= \frac{2e^{i\theta}}{1-\{1-2\sin^2(\pi/2n) + 2i\sin(\pi/2n)\cos(\pi/2n)\}}$$

$$= \frac{e^{i\theta}}{-i\sin(\pi/2n)[\cos(\pi/2n) + i\sin(\pi/2n)]}$$

$$= \frac{ie^{i\theta}(\pi/2n) - i\sin(\pi/2n)}{-i^2\sin(\pi/2n)}$$

$= i(\cos\theta + i\sin\theta)[\cos(\pi/2n) - i\sin(\pi/2n)]\operatorname{cosec}(\pi/2n).$

Equating the real parts on both the sides, we get

$C = [\cos\theta\sin(\pi/2n) - \sin\theta\cos(\pi/2n)]\operatorname{cosec}(\pi/2n).$

Hence $C = \sin\left(\frac{\pi}{2n} - \theta\right)\operatorname{cosec}\left(\frac{\pi}{2n}\right).$

Example 8:

Sum the series

(i) $\cos\theta + x\cos 2\theta + x^2\cos 3\theta + \ldots$ *n terms.*

(ii) $\sin\theta + x\sin 2\theta + x^2\sin 3\theta + \ldots$ *n terms.*

Solution:

Let $\quad C = \cos\theta + x\cos 2\theta + x^2\cos 3\theta + \ldots + x^{n-1}\cos n\theta.$

and $\quad S = \sin\theta + x\sin 2\theta + x^2\sin 3\theta + \ldots + x^{n-1}\sin n\theta.$

$$\therefore C + iS = e^{i\theta} + xe^{2i\theta} + x^2e^{3i\theta} + \ldots + x^{n-1}e^{ni\theta}$$

$$= e^{i\theta}\,[1 + xe^{i\theta} + x^2e^{3i\theta} + \ldots + x^{n-1}e^{(n-1)\theta}]$$

$$= e^{i\theta}\,(1 + z + z^2 + \ldots + z^{n-1}), \text{ where } z = xe^{i\theta}$$

$$= e^{i\theta}\left\{\frac{1-z^n}{1-z}\right\} = e^{i\theta}\left\{\frac{1-x^n e^{in\theta}}{1-xe^{i\theta}}\right\}.$$

$$= e^{i\theta}\left\{\frac{1-x^n e^{in\theta}}{1-xe^{i\theta}} \times \frac{1-xe^{-i\theta}}{1-xe^{-i\theta}}\right\}$$

$$= e^{i\theta}\left[\frac{1-xe^{-i\theta}-x^n e^{in\theta}+x^{n+1}e^{i(n-1)\theta}}{1-x\left(e^{i\theta}+e^{-i\theta}\right)+x^2}\right]$$

$$= \frac{e^{i\theta}-x-x^n\,e^{i(n+1)\theta}+x^{n+1}\,e^{in\theta}}{1-2x\cos\theta+x^2}$$

$$= \frac{\left(\cos\theta+i\sin\theta\right)-x-x^n\left\{\cos(n+1)\theta+i\sin(n+1)\theta\right\} + x^{n+1}\left\{\cos n\theta+i\sin n\theta\right\}}{1-2x\cos\theta+x^2}$$

Comparing the real and imaginary parts, we get

$$C = \frac{\cos\theta - x - x^n\cos(n+1)\theta + x^{n+1}\cos n\theta}{1-2x\cos\theta+x^2},$$

$$S = \frac{\sin\theta - x^n\sin\left(n+1\right)\theta + x^{n+1}\sin n\theta}{1-2x\cos\theta+x^2}.$$

Example 9:

Sum the n terms

(i) $\cos^2\theta + \cos^2 2\theta + \cos^2 3\theta + \ldots.$

(ii) $\sin^2 \theta + \sin^2 2\theta + \sin^2 3\theta + ...$

Solution:

(i) Let $C = \cos^2 \theta + \cos^2 2\theta + ... + \cos^2 n\theta$

$$= \frac{1}{2}(1 + \cos 2\theta) + \frac{1}{2}(1 + \cos 4\theta) + ... + \frac{1}{2}(1 + \cos 2n\theta)$$

$$= \frac{n}{2} + \frac{1}{2}(\cos 2\theta + \cos 4\theta + ... + \cos 2n\theta)$$

$$= \frac{n}{2} + \frac{1}{2}(\cos \phi + \cos 2\phi + ... + \cos n\phi), \ \phi = 2\theta$$

$$= \frac{n}{2} + \frac{1}{2}\left[\frac{\cos \phi - 1 - \cos(n+1)\phi + \cos n\phi}{2(1-\cos \phi)}\right]$$

$$\text{Hence } C = \frac{n}{2} + \frac{\cos 2\theta - 1 - \cos 2(n+1)\theta + \cos 2n\theta}{4(1-\cos 2\theta)}, \text{ if } \theta \neq 0$$

$= n$ if $\theta = 0$.

(ii) Let $S = \sin^2 \theta + \sin^2 2\theta + \sin^2 3\theta + ... + \sin^2 n\theta$.

$$= \frac{1}{2}(1 - \cos 2\theta) + \frac{1}{2}(1 - \cos 4\theta) + ... + \frac{1}{2}(1 - \cos 2n\theta)$$

$$= \frac{n}{2} - \frac{1}{2}(\cos 2\theta + \cos 4\theta + ... + \cos 2n\theta)$$

$$\text{Hence } S = \frac{n}{2} - \frac{\cos 2\theta - 1 - \cos 2(n+1)\theta + \cos 2n\theta}{4(1-\cos 2\theta)}, \text{ if } \theta \neq 0$$

$= 0$, if $\theta = 0$.

Example 10:

Sum the series

(i) $\cos \theta \cos \theta + \cos^2 \theta \cos 2\theta + \cos^3 \theta \cos 3\theta + ...$ *to n terms.*

(ii) $\cos \theta \sin \theta + \cos^2 \theta \sin 2\theta + \cos^3 \theta \sin 3\theta + ...$ *to n terms.*

Solution:

Let

$C = \cos \theta \cos \theta + \cos^2 \theta \cos 2\theta + ... + \cos^n \theta \cos n\theta$,

$S = \cos \theta \sin \theta + \cos^2 \theta \sin 2\theta + ... + \cos^n \theta \sin n\theta$.

$\therefore$ $C + iS = \cos\theta(\cos\theta + i\sin\theta) + \cos^2\theta(\cos 2\theta + i\sin 2\theta) +$
$+ \ldots + \cos^n\theta(\cos n\theta + i\sin n\theta)$

$= z\cos\theta + z^2\cos^2\theta + \ldots + z^n\cos^n\theta,$

where $z = \cos\theta + i\sin\theta$

$= z\cos\theta\,(1 + z\cos\theta + \ldots + z^{n-1}\cos^{n-1}\theta)$

$$= \frac{z\cos\theta\left[1 - z^n\cos^n\theta\right]}{1 - z\cos\theta}$$

$$= \frac{\cos\theta(\cos\theta + i\sin\theta)]\left[1 - \cos^n\theta\left\{\cos n\theta + i\sin n\theta\right\}\right]}{1 - \cos\theta\left\{\cos\theta + i\sin\theta\right\}}$$

$$= \frac{\cos\theta(\cos\theta + i\sin\theta)\left[1 - \cos^n\theta\cos n\theta - i\cos^n\theta\sin n\theta\right]}{(1 - \cos^2\theta) - i\sin\theta\cos\theta}$$

$$= \frac{\cos\theta(\cos\theta + i\sin\theta)\left[1 - \cos^n\theta\cos n\theta - i\cos^n\theta\sin n\theta\right]}{-i\sin\theta\,(\cos\theta + i\sin\theta)}$$

$= i\cot\theta\,[1 - \cos^n\theta\cos n\theta - i\cos^n\theta\sin n\theta]$

$= i\cot\theta\,[1 - \cos^n\theta\cos n\theta] + \cos^{n+1}\theta\,\mathrm{cosec}\,\theta\sin n\theta.$

Equating the real and imaginary parts on both the sides, we get

$C = \cos^{n+1}\theta\,\mathrm{cosec}\,\theta\sin n\theta$, if $\theta \neq 0$

and $C = n$, if $\theta = 0$;

$S = \cot\theta\,(1 - \cos^n\theta\cos n\theta)$, if $\theta \neq 0$

and $S = 0$, if $\theta = 0$.

Example 11:

Sum to n terms:

(i) $\cos\theta\cos 2\theta + \cos 2\theta\cos 3\theta + \cos 3\theta\cos 4\theta + \ldots$

(ii) $\sin\theta\sin 2\theta + \sin 2\theta\sin 3\theta + \sin 3\theta\sin 4\theta + \ldots$

Solution:

$C = \cos\theta\cos 2\theta + \cos 2\theta\cos 3\theta + \ldots + \cos n\theta\cos(n + 1)\theta$

$$= \frac{1}{2}(\cos 3\theta + \cos\theta) + \frac{1}{2}(\cos 5\theta + \cos\theta)$$
$$+ \ldots. + \frac{1}{2}[\cos(2n + 1)\theta + \cos\theta]$$

$$= \frac{1}{2} [\cos 3\theta + \cos 5\theta + ... + \cos (2n + 1)] + \frac{n}{2} \cos \theta$$

$$= \frac{1}{2} C_1 + \frac{n}{2} \cos \theta, \qquad ...(1)$$

where $C_1 \cos = 3\theta + \cos 5\theta + ... + \cos (2n + 1) \theta$.

Multiplying both sides by 2 sin θ, we get

$$2 \sin \theta\, C_1 = 2 \sin \theta \cos 3\theta + 2 \sin \theta \cos 5\theta + ... + 2 \sin \theta \cos (2n + 1) \theta$$

$$= (\sin 4\theta - \sin 2\theta) + (\sin 6\theta - \sin 4\theta) + ... + [\sin (2n + 2) \theta - \sin 2 n\theta]$$

$$= \sin (2n + 2) \theta - \sin 2\theta$$

$$= 2 \cos (n + 2) \theta \sin n\theta$$

$$\therefore C_1 = \frac{\cos (n+2) \theta \sin n\theta}{\sin \theta} \qquad ...(2)$$

Substituting the value of C_1 from (2) in (1), we obtain

$$C = \frac{\cos (n+2) \theta \sin n\theta}{2 \sin \theta} + \frac{n}{2} \cos \theta$$

$$= \frac{1}{2} [n \cos \theta + \cos (n + 2) \theta \sin n\theta \operatorname{cosec} \theta], \theta \neq k\pi.$$

Clearly C = n if θ is an even multiple of π,

= – n if θ is an odd multiple of π.

(ii) Let $S = \sin \theta \sin 2\theta \sin 3\theta + ... + \sin n\theta \sin (n + 1) \theta$

$$= \frac{1}{2} (\cos \theta - \cos 3\theta) + \frac{1}{2} (\cos \theta - \cos 5\theta) + ... + \frac{1}{2} [\cos \theta - \cos (2n + 1) \theta]$$

$$= \frac{n}{2} \cos \theta - \frac{1}{2} [\cos 3\theta + \cos 5\theta + ... + \cos (2n + 1) \theta]$$

$$= \frac{n}{2} \cos \theta - \frac{1}{2} C_1.$$

Hence, on using (2),

$$S = \frac{1}{2} [n \cos \theta - \cos (n + 2) \theta \sin n\theta \operatorname{cosec} \theta]\ \theta \neq k\pi$$

= 0, if θ = kπ.

Example 12:

Sum to n terms:

$$\sin\theta\cos 2\theta + \sin 2\theta\cos 3\theta + \sin 3\theta\cos 4\theta + \ldots$$

Solution:

Let $A = \sin\theta\cos 2\theta + \sin 2\theta\cos 3\theta + \ldots + \sin n\theta\cos(n+1)\theta$

$$= \frac{1}{2}(\sin 3\theta - \sin\theta) + \frac{1}{2}(\sin 5\theta - \sin\theta) + \ldots + \frac{1}{2}[\sin(2n+1)\theta - \sin\theta]$$

$$= \frac{1}{2}[\sin 3\theta + \sin 5\theta + \ldots + \sin(2n+1)\theta] - \frac{n}{2}\sin\theta$$

$$= \frac{1}{2}S - \frac{n}{2}\sin\theta. \qquad \ldots(3)$$

where $S = \sin 3\theta + \sin 5\theta + \ldots + \sin(2n+1)\theta$

Multiplying both sides by $2\sin\theta$, we get

$$2\sin\theta\, S = 2\sin\theta\sin 3\theta + 2\sin\theta\sin 5\theta + \ldots + 2\sin\theta\sin(2n+1)\theta$$

$$= (\cos 2\theta - \cos 4\theta) + (\cos 4\theta - \cos 6\theta) + \ldots + [\cos 2n\theta - \cos(2n+2)\theta]$$

$$= \cos 2\theta - \cos(2n+2)\theta$$

$$= 2\sin(n+2)\theta\sin n\theta$$

$\therefore S = \sin(n+2)\theta\sin n\theta\operatorname{cosec}\theta.$

Substituting for S in (3), we get

$$A = \frac{1}{2}[\sin(n+2)\theta\sin n\theta\operatorname{cosec}\theta - n\sin\theta], \text{ if } \theta \neq k\pi$$

$$= 0, \text{ if } \theta = k\pi.$$

Example 13:

Sum the series

(i) $\cos\theta + \cos 3\theta + \cos 5\theta + \ldots$ *to n terms.*

(ii) $\sin\theta + \sin 3\theta + \sin 5\theta + \ldots$ *to n terms.*

Solution:

(i) Let $C = \cos\theta + \cos 3\theta + \cos 5\theta + \ldots + \cos(2n-1)\theta.$

$\therefore$ 2 sin θ C = 2 sin θ cos θ + 2 sin θ cos 3θ + 2 sin θ cos 5θ + ...
+ 2 sin θ cos (2n – 1) θ

= sin 2θ + (sin 4θ – sin 2θ) + (sin 6θ – sin 4θ) + ...
+ [sin 2nθ – sin (2n ...2) θ]

= sin 2nθ.

Hence $C = \dfrac{\sin 2n\theta}{2 \sin \theta}$.

(ii) Let S = sin θ + sin 3θ + sin 5θ + ... + sin (2n – 1) θ

$\therefore$ 2 sin θ S = $2 \sin^2 \theta$ + 2 sin 3θ sin θ + 2 sin 5θ sin θ + ...
+ 2 sin (2n – 1) θ sin θ

= $2 \sin^2 \theta$ + (cos 2θ – cos 4θ) + (cos 4θ – cos 6θ) + ...
+ [cos (2n – 2) θ – cos 2nθ}

= $2 \sin^2 \theta$ + cos 2θ – cos 2nθ

= $2 \sin^2 \theta + 2 \cos^2 \theta - 1 - \{1 - 2 \sin^2 n\theta\}$

= $2 \sin^2 n\theta$.

Hence $S = \dfrac{\sin^2 n\theta}{\sin \theta}$.

Example 14:

Sum the series

$$\cos \alpha + n \cos 2\alpha + \frac{n(n-1)}{2!} \cos 3\alpha + \dots \text{ upto } (n+1) \text{ terms.}$$

Solution:

Let $C = \cos \alpha + n \cos 2\alpha + \dfrac{n(n-1)}{2!} \cos 3\alpha + \dots$

and $S = \sin \alpha + n \sin 2\alpha + \dfrac{n(n-1)}{2!} \sin 3\alpha + \dots$

$\therefore C + iS = (\cos \alpha + i \sin \alpha) + n (\cos 2\alpha + i \sin 2\alpha) + \dots$

$= e^{i\alpha} + ne^{i2\alpha} + \dfrac{n(n-1)}{2!} e^{i2\alpha} + \dots$ to (n + 1) terms

$\Rightarrow C + iS = e^{i\alpha} \left[1 + ne^{i\alpha} + \dfrac{n(n-1)}{2!} e^{i2\alpha} + \dots e^{in\alpha} \right]$

$= e^{i\alpha} (1 + e^{i\alpha})^n$

$$= (\cos\alpha + i\sin\alpha)(1 + \cos\alpha + i\sin\alpha)^n$$

$$= (\cos\alpha + i\sin\alpha)\left(2\cos^2\frac{\alpha}{2} + 2i\sin\frac{\alpha}{2}\cos\frac{\alpha}{2}\right)^n$$

$$= 2^n\cos^n\frac{\alpha}{2}(\cos\alpha + i\sin\alpha)\left(\cos\frac{\alpha}{2} + i\sin\frac{\alpha}{2}\right)^n$$

$$= 2^n\cos^n\frac{\alpha}{2}(\cos\alpha + i\sin\alpha)\left(\cos\frac{n\alpha}{2} + i\sin\frac{n\alpha}{2}\right)$$

Comparing the real parts on both the sides, we get

$$C = 2^n\cos^n\frac{\alpha}{2}\left[\cos\alpha\cos\frac{n\alpha}{2} - \sin\alpha\sin\frac{n\alpha}{2}\right]$$

$$= 2^n\cos^n\frac{\alpha}{2}\cos\left(\alpha + \frac{n\alpha}{2}\right)$$

Hence $C = 2^n\cos^n\dfrac{\alpha}{2}\cos\dfrac{(n+2)\alpha}{2}$.

Example 15:

Apply Summation of Series to prove that the equation whose roots are nth powers of the roots of the the equation $x^2 - 2\cos\theta\, x + 1 = 0$ is $x^2 - 2\cos n\theta\, x + 1 = 0$.

Solution:

Solving the equation $x^2 - 2\cos\theta\, x + 1 = 0$, we get

$$x = \frac{2\cos\theta \pm \sqrt{4\cos^2\theta - 4}}{2} = \cos\theta \pm i\sin\theta.$$

The nth powers of the above two roots are

$$\alpha = (\cos\theta + i\sin\theta)^n,\ \beta = (\cos\theta - i\sin\theta)^n$$

$\Rightarrow \alpha = \cos n\theta + i\sin n\theta,\ \beta = \cos n\theta - i\sin n\theta$, by DeMoivre's theorem

$\therefore \alpha + \beta = 2\cos n\theta,\ \alpha\beta = \cos^2 n\theta + \sin^2 n\theta = 1.$

Hence the equation with roots α and β is

$$x^2 - 2(\alpha + \beta)x + \alpha\beta = 0$$

$$\Rightarrow \quad x^2 - 2\cos n\theta\, x + 1 = 0$$

Example 16:

If α, β be the roots of $x^2 - 2x + 2 = 0$, prove that $\alpha^n + \beta^n = 2^{\frac{n}{2}+1}$ cos $\left(\frac{n\pi}{4}\right)$ and hence evaluate $\alpha^6 + \beta^6$.

Solution:

Putting n = 6, we get:

$$\alpha^6 + \beta^6 = 2^4 \cos\left(\frac{6\pi}{4}\right) = 16 \cos\left(\frac{3\pi}{2}\right) = 0.$$

Example 17:

If α, β be the roots of $x^2 - 2x + 4 = 0$, prove that $\alpha^n + \beta^n = 2^{n-1}$ cos $\frac{n\pi}{3}$ and hence evaluate $\alpha^6 + \beta^6$.

Solution:

Solving $x^2 - 2x + 4 = 0$, we get

$$x = \frac{2 \pm \sqrt{4-16}}{2} = 1 \pm i\sqrt{3}.$$

Let $1 = r \cos\theta$ and $\sqrt{3} = r \sin\theta$. Then $r = 2$, $\tan\theta = \sqrt{3}$ or $\theta = \pi/3$.

We have $\alpha = r(\cos\theta + i \sin\theta)$, $\beta = r(\cos\theta - i\sin\theta)$.

$\therefore \alpha^n = r^n(\cos n\theta + i \sin n\theta)$, $\beta^n = r^n(\cos n\theta - i \sin n\theta)$

Hence $\alpha^n + \beta^n = 2r^n \cos n\theta = 2.2^n \cos\frac{n\pi}{3} = 2^{n+1} \cos\frac{n\pi}{3}$.

Putting n = 6, we get $\alpha^6 + \beta^6 = 2^7 \cos 2\pi = 128$.

Example 18:

If $\cos 2\theta + i \sin 2\theta = p$, $\cos 2\phi + i \sin 2\phi = q$, prove that

$$2\cos(\theta - \phi) = \sqrt{\frac{p}{q}} + \sqrt{\frac{q}{p}},\quad 2i \sin(\theta - \phi) = \sqrt{\frac{p}{q}} - \sqrt{\frac{q}{p}}.$$

Solution:

We have $\sqrt{p} = (\cos 2\theta + i \sin 2\theta)^{1/2} = \cos\theta + i\sin\theta$.

Similarly, $\sqrt{q} = \cos\phi + i \sin\phi$.

$$\therefore \quad \sqrt{\frac{p}{q}} = \frac{\cos\theta + i\sin\theta}{\cos\phi + i\sin\phi} = \cos\ (\theta - \phi) + i\sin\ (\theta - \phi),$$

$$\text{and} \quad \sqrt{\frac{q}{p}} = \cos(\phi - \theta) + i\sin(\phi - \theta) = \cos(\theta - \phi) - i\sin(\theta - \phi).$$

$$\text{Hence} \quad \sqrt{\frac{p}{q}} + \sqrt{\frac{q}{p}} = 2\cos\ (\theta - \phi), \quad \sqrt{\frac{p}{q}} - \sqrt{\frac{q}{p}} = 2i\sin(\theta - \phi).$$

Example 19:

If $(1 + x)^n = c_0 + c_1x + ... + c_nx^n$, *n being a positive integer, prove that:*

(i) $c_0 - c_2 + c_4 - c_6 + ... = 2^{n/2} \cos n\pi/4,$

(ii) $c_1 - c_3 + c_5 - ... = 2^{n/2} \sin n\pi/4,$

(iii) $c_0 + c_4 + c_8 + ... = 2^{n-2} + 2^{(n-2)/2} \cos n\pi/4.$

Solution:

We have

$$(1 + x)^n = c_0 + c_1x + c_2x^2 + c_3x^3 + c_4x^4 + ...$$

Putting n = 1, – 1, i, – i successively, we get

$$2^n = c_0 + c_1 + c_2 + c_3 + c_4 + ...$$

$$0 = c_0 - c_1 + c_2 - c_3 + c_4 - ...$$

$$(1 + i)^n = c_0 + c_1i - c_2 - c_3i + c_4 ...$$

$$(1 - i)^n = c_0 - c_1i - c_2 + c_3i + c_4 ...$$

Adding the above relations, we get

$$2^n + (1 + i)^n + (1 - i)^n = 4\ (c_0 + c_4 + c_8 + ...)$$

$$\Rightarrow 2^n + 2^{n/2}\left(\cos\frac{n\pi}{4} + i\sin\frac{n\pi}{4}\right) + 2^{n/2}\left(\cos\frac{n\pi}{4} - i\sin\frac{n\pi}{4}\right)$$

$$= 4\ (c_0 + c_4 + c_8 + ...)$$

$$\Rightarrow 2^n + 2.2^{n/2}\cos\frac{n\pi}{4} = 2^2\ (c_0 + c_4 + c_8 + ...)$$

Dividing by 2^2 on both sides, we get:

$$c_0 + c_4 + c_8 + ... = 2^{n-2} + 2^{\frac{n-2}{2}}\cos\frac{n\pi}{4}.$$

Example 20:

If $(\cos\theta + i\sin\theta)(\cos 2\theta + i\sin 2\theta) ... (\cos n\theta + i\sin n\theta) = 1$, *prove that*

$$\theta = \frac{4k\pi}{n(n+1)}, \text{ where } k \text{ is any integer.}$$

Solution:

We have

$$(\cos\theta + i\sin\theta)(\cos 2\theta + i\sin 2\theta) \ldots (\cos n\theta + i\sin n\theta)$$
$$= (\theta + 2\theta + \ldots + n\theta) + i\sin(\theta + 2\theta + \ldots + n\theta)$$
$$= \cos(1 + 2 + \ldots + n)\theta + i\sin(1 + 2 + \ldots + n)\theta$$
$$= \cos\frac{n(n+1)\theta}{2} + i\sin\frac{n(n+1)\theta}{2}$$
$$= \cos 2k\pi + i\sin 2k\pi, \text{ where } 2k\pi = \frac{n(n+1)\theta}{2}$$
$$= 1, \text{ where } \theta = \frac{4k\pi}{n(n+1)}, \text{ k being any integer.}$$

Example 21:

Prove that:

$$\left(\frac{-1 + i\sqrt{3}}{2}\right) + \left(\frac{-1 - i\sqrt{3}}{2}\right)^n$$

has value 2 if n = 3k and – 1 if n = 3k ± 1, where k is an integer.

Solution:

Let $r\cos\theta = -\frac{1}{2}$ and $r\sin\theta = \frac{\sqrt{3}}{2}$. Then $r^2 = 1 \Rightarrow r = 1$, $\tan\theta = -\sqrt{3} \Rightarrow \theta = 2\pi/3$. We have

$$\left(\frac{-1 + i\sqrt{3}}{2}\right)^n + \left(\frac{-1 - i\sqrt{3}}{2}\right)^n = r^n(\cos\theta + i\sin\theta)^n + r^n(\cos\theta - i\sin\theta)^n$$

$$= \cos n\theta + i\sin n\theta + \cos n\theta - i\sin n\theta \qquad (\because r = 1)$$
$$= 2\cos n\theta = 2\cos\left(\frac{2n\pi}{3}\right)$$
$$= 2\cos 2k\pi, \text{ if } n = 3k \text{ (k any integer)}$$
$$= 2.1 = 2, \text{ if } n = 3k \text{ (k any integer)}$$

If $n = 3k \pm 1$, then

$$2\cos\frac{2n\pi}{3} = 2\cos\left(2k\pi \pm \frac{2\pi}{3}\right) = 2\cos\frac{2\pi}{3} = -2.\frac{1}{2} = -1.$$

Example 22:

From the identity:

$$\frac{1}{x-a} - \frac{1}{x-b} \equiv \frac{a-b}{(x-a)(x-b)}, \text{ prove that}$$

(i) $\cos(\theta+\alpha)\sin(\theta-\beta) - \cos(\theta+\beta)\sin(\theta-\alpha) = \sin(\alpha-\beta)\cos 2\theta,$

(ii) $\sin(\theta+\alpha)\sin(\theta-\beta) - \sin(\theta+\beta)\sin(\theta-\alpha)$
$= \sin(\alpha-\beta)\sin 2\theta.$

Solution:

Let $x = \cos 2\theta + i\sin 2\theta$, $a = \cos 2\alpha + i\sin 2\alpha$,

$b = \cos 2\beta + i\sin 2\beta$.

$\therefore\ x - a = (\cos 2\theta - \cos 2\alpha) + i(\sin 2\theta - \sin 2\alpha)$

$= -2\sin(\theta+\alpha)\sin(\theta-\alpha) + 2i\cos(\theta+\alpha)\sin(\theta-\alpha)$

$= 2i\sin(\theta-\alpha)\{\cos(\theta+\alpha) + i\sin(\theta+\alpha)\}.$

Similarly,

$x - b = 2i\sin(\theta-\beta)\{\cos(\theta+\beta) + i\sin(\theta+\beta)\},$

$a - b = 2i\sin(\alpha-\beta)\{\cos(\alpha+\beta) + i\sin(\alpha+\beta)\}.$

Putting these values in the given idenity, we get

$$\frac{1}{2i\sin(\theta-\alpha)}[\cos(\theta+\alpha) - i\sin(\theta+\alpha)]$$

$$-\frac{1}{2i\sin(\theta-\beta)}[\cos(\theta+\beta) - i\sin(\theta+\beta)]$$

$$= \frac{2i\sin(\alpha-\beta)[\cos(\alpha+\beta) + i\sin(\alpha+\beta)]}{-4\sin(\theta-\alpha)\sin(\theta-\beta)\{\cos(2\theta+\alpha+\beta) + i\sin(2\theta+\alpha+\beta)\}}$$

Multiplying both sides by 2i, we get:

$$= \frac{\cos(\theta+\alpha) - i\sin(\theta+\alpha)}{\sin(\theta-\alpha)} - \frac{\cos(\theta-\beta) - i\sin(\theta+\beta)}{\sin(\theta-\beta)}$$

$$= \frac{\sin(\alpha-\beta)\{\cos(\alpha+\beta-2\theta-\alpha-\beta) + i\sin(\alpha+\beta-2\theta-\alpha-\beta)\}}{\sin(\theta-\alpha)\sin(\theta-\beta)}$$

$$= \frac{\sin(\alpha-\beta)(\cos 2\theta - i\sin 2\theta)}{\sin(\theta-\alpha)\sin(\theta-\beta)}$$

Comparing the real parts on both the sides, we get

$$\frac{\cos(\theta+\alpha)}{\sin(\theta-\alpha)} - \frac{\cos(\theta+\beta)}{\sin(\theta-\beta)} = \frac{\sin(\alpha-\beta)\cos 2\theta}{\sin(\theta-\alpha)\sin(\theta-\beta)}$$

$\Rightarrow \cos(\theta+\alpha)\sin(\theta-\beta) - \cos(\theta+\beta)\sin(\theta-\alpha) = \sin(\alpha-\beta)\sin 2\theta$.

Similarly on comparing the imaginary parts, we get part (ii).

Example 23:

State Summation of Series for rational index and use it to solve the equation $z^7 + z = 0$.

Solution:

$Z^7 + z = 0 \Rightarrow z(z^6 + 1) = 0 \Rightarrow z = 0 \Rightarrow z^6 + 1 = 0$.

The six roots of $z^6 + 1 = 0$ are

$$+ i, \frac{1}{2}\left(\sqrt{3} \pm i\right), \frac{1}{2}\left(-\sqrt{3} \pm i\right).$$

Example 24:

Express $\cos 7\theta$, $\sin 7\theta$ in powers of $\cos\theta$ and $\sin\theta$, and hence obtain an expression for $\tan 7\theta$ in powers of $\tan\theta$.

Solution:

We have

$\cos 7\theta = \cos^7\theta - 21\cos^5\theta\sin^2\theta + 35\cos^3\theta\sin^4\theta - 7\cos\theta\sin^6\theta$,

$\sin 7\theta = 7\cos^6\theta\sin\theta - 35\cos^4\theta\sin^3\theta + 21\cos^2\theta\sin^5\theta - \sin^7\theta$.

Hence

$$\tan 7\theta = \frac{7\cos^6\theta\sin\theta - 35\cos^4\theta\sin^3\theta + 21\cos^2\theta\sin^5\theta - \sin^7\theta}{\cos^7\theta - 21\cos^5\theta\sin^2\theta + 35\cos^3\theta\sin^4\theta - 7\cos\theta\sin^6\theta}$$

On dividing numerator and denominator by $\cos^7\theta$, we get

$$\tan 7\theta = \frac{7\tan\theta - 35\tan^3\theta + 21\tan^5\theta - \tan^7\theta}{1 - 21\tan^2\theta + 35\tan^4\theta - 7\tan^6\theta}.$$

Example 25:

Show that

$$2^{11}\cos^7\theta\sin^5\theta = \sin 12\theta + 2\sin 10\theta - 4\sin 8\theta - 10\sin 6\theta + 5\sin 4\theta + 20\sin 2\theta.$$

Solution:

Let $z = \cos\theta + i\sin\theta$. Then $1/z = \cos\theta - i\sin\theta$, $z + 1/z = 2\cos\theta$, $z - 1/z = 2i\sin\theta$.

$$\therefore (2\cos\theta)^7(2i\sin\theta)^5 = \left(z+\frac{1}{z}\right)^7\left(z-\frac{1}{z}\right)^5$$

$$= \left(z+\frac{1}{z}\right)^2\left(z^2-\frac{1}{z^2}\right)^5$$

$$= (z^2 + z^{-2} + 2)(z^{10} - 5z^6 + 10z^2 - 10z^{-2} + 5z^{-6} - z^{-10})$$

$$= (z^{12} - z^{-12}) + 2(z^{10} - z^{-10}) - 4(z^8 - z^{-8}) - 10(z^6 - z^{-6}) + 5(z^4 - z^{-4}) + 20(z^2 - z^{-2})$$

$$\Rightarrow 2^{12}.i.\cos^7\theta\sin^5\theta = 2i\sin 12\theta + 2(2i\sin 10\theta) - 4(2i\sin 8\theta) - 10(2i\sin 6\theta) + 5(2i\sin 4\theta) + 20(2i\sin 2\theta).$$

Hence $2^{11}\cos^7\theta\sin^5\theta = \sin 12\theta + 2\sin 10\theta - 4\sin 8\theta - 10\sin 6\theta + 5\sin 4\theta + 20\sin 2\theta$,

Example 26:

Find the sum to infinity of the series:

$$\sin\alpha + \frac{1}{2}\sin 3\alpha + \frac{1.3}{2.4}\sin 5\alpha + \ldots,\ -\pi < \alpha < \pi.$$

Solution:

$$\text{Let } S = \sin\alpha + \frac{1}{2}\sin 3\alpha + \frac{1.3}{2.4}\sin 5\alpha + \ldots$$

and $$C = \cos\alpha + \frac{1}{2}\cos 3\alpha + \frac{1.3}{2.4}\cos 5\alpha + \ldots$$

$$\therefore C + iS = e^{i\alpha} + \frac{1}{2}e^{3i\alpha} + \frac{1.3}{2.4}e^{5i\alpha} + \ldots$$

$$= e^{i\alpha}\left[1+\frac{1}{2}e^{2i\alpha}+\frac{1.3}{2.4}e^{4i\alpha}+\ldots\right]$$

$$= e^{i\alpha}(1 - e^{2i\alpha})^{-1/2}$$

$$= \frac{\cos\alpha + i\sin\alpha}{[1-(\cos 2\alpha + i\sin 2\alpha)]^{1/2}}$$

$$= \frac{\cos\alpha + i\sin\alpha}{\left(2\sin^2\alpha - 2i\sin\alpha\cos\alpha\right)^{1/2}}$$

$$= \frac{\cos\alpha + i\sin\alpha}{\sqrt{2\sin\alpha}\left(\sin\alpha - i\cos\alpha\right)^{1/2}}$$

$$= \frac{\cos\alpha + i\sin\alpha}{\sqrt{2\sin\alpha}\left\{\cos\left(\pi/2-\alpha\right) - i\sin\left(\pi/2-\alpha\right)\right\}^{1/2}}$$

$$= \frac{\cos\alpha + i\sin\alpha}{\sqrt{2\sin\alpha}\left\{\cos\left(\pi/4-\alpha/2\right) - i\sin\left(\pi/4-\alpha/2\right)\right\}}$$

by DeMoivre's theorem

$$= \frac{1}{\sqrt{2\sin\alpha}}\left\{\cos\left(\alpha+\frac{\pi}{4}-\frac{\alpha}{2}\right) + i\sin\left(\alpha+\frac{\pi}{4}-\frac{\alpha}{2}\right)\right\}$$

$$= \frac{1}{\sqrt{2\sin\alpha}}\left\{\cos\left(\frac{\pi}{4}+\frac{\alpha}{2}\right) + i\sin\left(\frac{\pi}{4}+\frac{\alpha}{2}\right)\right\}. \qquad ...(1)$$

Equating the imaginary parts on both sides, we get

$$S = \sin\left(\frac{\pi}{4}+\frac{\alpha}{2}\right)\ (2\sin a)^{-1/2}.$$

Example 27:

Find the sum to infinity of the series

$$C: 1 + \frac{1}{2}\cos 2\theta - \frac{1}{2.4}\cos 4\theta + \frac{1.3}{2.4.6}\cos 6\theta - ...$$

Solution:

$$\text{Let } S = \frac{1}{2}\sin 2\theta - \frac{1}{2.4}\sin 4\theta + \frac{1.3}{2.4.6}\cos 6\theta - ...$$

$$\text{Then}\quad C + iS = 1 + \frac{1}{2}e^{2i\theta} - \frac{1}{2.4}e^{4i\theta} + \frac{1.3}{2.4.6}e^{6i\theta} - ...$$

$$= (1 + e^{2i\theta})^{1/2} = = (1 + \cos 2\theta + i\sin 2\theta)^{1/2}$$

$$= (2\cos^2\theta + 2i\sin\theta\cos\theta)^{1/2}$$

$$= \sqrt{2\cos\theta}\ (\cos\theta/2 + i\sin\theta/2)$$

Hence $C = \sqrt{2\cos\theta}\ \cos\theta/2$.

Example 28:

Sun the series: $\sin\theta\sin\theta + \sin^2\theta\sin 2\theta + ... + \sin^n\theta\sin n\theta$.

Solution:

Let $S = \sin\theta \sin\theta + \sin^2\theta \sin 2\theta + ... + \sin^n\theta \sin n\theta$,

$C = \sin\theta \cos\theta + \sin^2\theta \cos 2\theta + ... + \sin^n\theta \cos n\theta$.

$\therefore C + iS = z \sin\theta + z^2 \sin^2\theta + ... + z^n \sin^n\theta,\ z = \cos\theta + i\sin\theta$

$= z \sin\theta\,(1 + z\sin\theta + ... + z^{n-1}\sin^{n-1}\theta)$

$$= \frac{z\sin\theta\left(1 - z^n \sin^n\theta\right)}{1 - z\cos\theta}$$

$$= \frac{\sin\theta\left(\cos\theta + i\sin\theta\right)\left[1 - \sin^n\theta\left(\cos n\theta + i\sin n\theta\right)\right]}{1 - \cos\theta\left(\cos\theta + i\sin\theta\right)}$$

$$= \frac{\sin\theta\left(\cos\theta + i\sin\theta\right)\left[\left(1 - \sin^n\theta\cos n\theta\right) - i\sin^n\theta\sin n\theta\right]}{\sin^2\theta - i\sin\theta\cos\theta = -i\sin\theta\left(\cos\theta + i\sin\theta\right)}$$

$= i\,[(1 - \sin^n\theta\cos n\theta) - i\sin^n\theta\sin n\theta]$. ...(1)

Comparing the imaginary parts on both sides, we get

$S = 1 - \sin^n\theta\cos n\theta$.

Example 29:

Sum the series

$3\sin\alpha + 5\sin 2\alpha + 7\sin 3\alpha + ...$ n terms.

Solution:

Let $S = 3\sin\alpha + 5\sin 2\alpha + 7\sin 3\alpha + ... + (2n + 1)\sin n\alpha$

and $C = 1 + 3\cos\alpha + 5\cos 2\alpha + 7\cos 3\alpha + ... + (2n + 1)\cos n\alpha$.

$\therefore C + iS = 1 + 3e^{ia} + 5e^{2i\alpha} + 7e^{3i\alpha} + ... + (2n + 1)\,e^{ni\alpha}$....(1)

Multiplying both sides by $e^{i\alpha}$, we get

$(C + iS)\,e^{i\alpha} = e^{i\alpha} + 3e^{2i\alpha} + 5e^{4i\alpha} + ... + (2n + 1)\,e^{(n+1)i\alpha}$. ...(2)

Subtracting (2) from (1), we get

$(C + iS)(1 - e^{i\alpha}) = 1 + 2\,(e^{i\alpha} + e^{2i\alpha} + ... + e^{ni\alpha}) - (2n + 1)\,e^{(n+1)i\alpha}$

$= 1 + 2e^{i\alpha}\,\{1 + e^{i\alpha} + e^{2i\alpha} + ...\ e^{(n-1)\,i\alpha}\} - (2n + 1)\,e^{(n+1)i\alpha}$

$$= 1 + 2e^{i\alpha}\left[\frac{1 - e^{ni\alpha}}{\cdot 1 - e^{i\alpha}}\right] - (2n + 1)\,e^{(n+1)}\,i\alpha,\ \text{if } e^{i\alpha} \neq 1$$

$$= \frac{1 + e^{i\alpha} - 2e^{(n+1)i\alpha} - \left(2n + 1\right)e^{(n+1)i\alpha} + \left(2n + 1\right)e^{(n+2)\,i\alpha}}{1 - e^{i\alpha}}$$

$$\therefore C + iS = \frac{1+e^{i\alpha} - (2n+3)\, e^{(n+1)i\alpha} + (2n+1)\, e^{(n+1)i\alpha}}{\left(1-e^{i\alpha}\right)^2} \times \frac{e^{-i\alpha}}{e^{-i\alpha}}$$

$$= \frac{e^{-i\alpha} + 1 - (2n+3)\, e^{ni\alpha} + (2n+1)\, e^{(n+1)i\alpha}}{\left\{e^{-i\alpha/2}\left(1-e^{i\alpha}\right)\right\}^2 = \left(e^{-i\alpha/2} - e^{i\alpha/2}\right)^2}$$

$$= \frac{\cos\alpha - i\sin\alpha + 1 - (2n+3)(\cos n\alpha + i\sin n\alpha) + (2n+1)\{\cos(n+1)\alpha + i\sin(n+1)\alpha\}}{(2i\sin\alpha/2)^2 = -4\sin^2\alpha/2 = -2(1-\cos\alpha)}$$

Comparing imaginary parts on both sides, we get

$$S = \frac{\sin\alpha + (2n+3)\sin\alpha - (2n+1)\sin(n+1)\alpha}{2(1-\cos\alpha)}$$

if $\alpha \neq 2k\pi$ (k being any integer).

[$\because$ $e^{i\alpha} = 1 \Rightarrow \cos\alpha + i\sin\alpha = 1 \Rightarrow \cos\alpha = 1$ and $\sin\alpha = 0 \Rightarrow \alpha = 2k\pi$, k being any integer]

If $\alpha = 2k\pi$, the sum of the given series is obviously zero.

Example 30:

Sum the series to n terms:

(i) $\sin\alpha + 3\sin 2\alpha + 5\sin 3\alpha + \ldots$

(ii) $\cos\alpha + 3\cos 2\alpha + 5\cos 3\alpha + \ldots$

Solution:

Let $S = \sin\alpha + 3\sin 2\alpha + 5\sin 3\alpha + \ldots + (2n-1)\sin n\alpha$

and $C = \cos\alpha + 3\cos 2\alpha + 5\cos 3\alpha + \ldots + (2n-1)\cos n\alpha.$

$\therefore C + iS = e^{i\alpha} + 3e^{2i\alpha} + 5e^{3i\alpha} + \ldots + (2n-1)\, e^{ni\alpha}.$ ...(1)

Multiplying both sides by $e^{i\alpha}$, we get:

$(C + iS)\, e^{i\alpha} = e^{2i\alpha} + 3e^{3i\alpha} + 5e^{4i\alpha} + \ldots + (2n-1)\, e^{(n+1)i\alpha}.$ ...(2)

Subtracting (2) from (1), we get

$(C + iS)(1 - e^{i\alpha}) = e^{i\alpha} + 2e^{3i\alpha} + 2e^{3i\alpha} + \ldots + 2e^{ni\alpha} - (2n-1)\, e^{(n+1)i\alpha}$

$= e^{i\alpha} + 2e^{2i\alpha}\{1 + e^{i\alpha} + e^{2i\alpha} + \ldots + e^{(n-2)i\alpha}\} - (2n-1)\, e^{(n+1)i\alpha}$

$$= e^{i\alpha} + 2e^{i\alpha}\left[\frac{1-e^{(n-1)i\alpha}}{1-e^{i\alpha}}\right] - (2n-1)\, e^{(n+1)i\alpha}, \text{ if } e^{i\alpha} \neq 1$$

$$= \frac{e^{i\alpha} + e^{2i\alpha} - 2e^{(n+1)\,i\alpha} - (2n-1)\, e^{(n+1)\,i\alpha} + (2n-1)\, e^{(n+2)\,i\alpha}}{(1-e^{i\alpha})}$$

We have

$$C + iS = \frac{e^{i\alpha} + e^{2i\alpha} - (2n+1) e^{(n+1)\,i\alpha} + (2n-1)\, e^{(n+2)\,i\alpha}}{(1-e^{i\alpha})^2} \times \frac{e^{-i\alpha}}{e^{-i\alpha}}$$

$$= \frac{1 + e^{i\alpha} - (2n+1)\, e^{ni\alpha} + (2n-1)\, e^{(n+1)\,i\alpha}}{(e^{-i\alpha/2} - e^{i\alpha/2})^2 = (2i \sin \alpha/2)^2 = -4 \sin^2 \alpha/2}$$

$$= \frac{1 + (\cos \alpha + i \sin \alpha) - (2n+1)(\cos n\alpha + i \sin n\alpha) + (2n-1)\{\cos (n+1)\alpha + i \sin (n+1)\alpha\}}{-2(1-\cos \alpha)} \quad ...(1)$$

Comparing the imaginary parts on both the sides, we get

$$S = \frac{(2n+1) \sin n\alpha - (2n-1) \sin (n+1)\alpha - \sin \alpha}{2(1-\cos \alpha)},$$

provided $e^{i\alpha} \neq 1$ *i.e.*, $\alpha \neq 2k\pi$, k being any integer.

[$\because e^{i\alpha} = 1 \Rightarrow \cos \alpha + i \sin \alpha = 1 \Rightarrow \cos \alpha = 1$ and $\sin \alpha = 0 \Rightarrow \alpha = 2k\pi$, k being any integer.]

If $\alpha = 2k\pi$, the sum of the series is obviously zero.

(ii) Comparing the real parts on both the sides of (1), we get

$$C = \frac{(2n+1) \cos n\alpha - (2n-1) \cos (n+1)\alpha - \cos \alpha - 1}{2(1-\cos \alpha)}, \ a \neq 2k\pi$$

If $\alpha = 2k\pi$, the sum of the given series is

$1 + 3 + 5 + ... + (2n - 1) = n^2$.

Example 31:

Sum the the series to n terms:

$\cos \alpha + 2 \cos (\alpha + \beta) + 3 \cos (\alpha + 2\beta) + ...$

Solution:

Let

$$C = \cos \alpha + 2 \cos (\alpha + \beta) + 3 \cos (\alpha + 2\beta) + ... + n \cos \left(a + \overline{n-1}\, \beta\right)$$

$S = \sin\alpha + 2\sin(\alpha+\beta) + 3\sin(\alpha+2\beta) + ... + n\sin\left(a+\overline{n-1}\,\beta\right)$

$\therefore\ C + iS = e^{i\alpha} + 2e^{i(\alpha+\beta)} + 3e^{i(\alpha+2\beta)} + ... + ne^{i(a+\overline{n-1}\,\beta)}$. ...(1)

Multiplying (2) from (1), we get

$(C + iS)(1 - e^{i\beta}) = e^{i\alpha} + e^{i(\alpha+\beta)} + e^{i(\alpha+3\beta)} + ... + ne^{i(\alpha+n\beta)}$ ...(2)

Subtracting (2) from (1), we get

$(C + iS)(1 - e^{i\beta}) = e^{i\alpha} + e^{i(\alpha+\beta)} + e^{i(\alpha+2\beta)} + ...\, e^{i(\varepsilon+\overline{n-1}\,\beta)} - ne^{i(\alpha+n\beta)}$

$= e^{i\alpha}[1 + e^{i\beta} + e^{2i\beta} + ... + e^{(n-1)i\beta}] - ne^{i(\alpha+n\beta)}$

$= e^{i\alpha}\left(\dfrac{1-e^{ni\beta}}{1-e^{i\beta}}\right) - ne^{i(a+nb)}$, provided $e^{i\beta} \neq 1$

$$= \frac{e^{i\alpha} - (n+1)\,e^{i(\alpha+n\beta)} + ne^{i(\alpha+\overline{n+1}\,\beta)}}{1-e^{i\beta}}$$

We have

$$C + iS = \frac{e^{i\alpha} - (n+1)\,e^{i(\alpha+n\beta)} + e^{i(\alpha+\overline{n+1}\,\beta)}}{(1-e^{i\beta})^2} \times \frac{\alpha^{-i\beta}}{e^{-i\beta}}$$

$$= \frac{e^{i(\alpha-\beta)} - (n+1)\,e^{i(\alpha+\overline{n-1}\,\beta)} + ne^{i(\alpha+n\beta)}}{(e^{-i\beta/2} - e^{i\beta/2})^2 = (2i\sin\beta/2)^2 = -4\sin^2\beta/2}$$

Comparing the real parts on both the sides, we get

$C = \dfrac{1}{4}[(n+1)\{\cos\alpha + (n-1)\beta\} - n\cos n(\alpha+n\beta) - \cos(\alpha-\beta)]$

$\times \operatorname{cosec}^2\beta/2$,

if $e^{i\beta} \neq 1$ *i.e.*, $\beta \neq 2k\pi$ (k being any integer).

If $\beta = 2k\pi$, the sum of the given series is

$\cos\alpha + 2\cos\alpha + 3\cos\alpha + ... + n\cos\alpha = \dfrac{n(n+1)}{2}\cos\alpha$.

A Useful Formula

$[x - (\cos\alpha + i\sin\alpha)][x - (\cos\alpha - i\sin\alpha)] = x^2 - 2x\cos\alpha + 1$.

To verify this, we see that

L.H.S. $= [(x - \cos\alpha) - i\sin\alpha][(x - \cos\alpha) + i\sin\alpha]$

$= (x - \cos\alpha)^2 - (i\sin\alpha)^2 = (x - \cos\alpha)^2 + \sin^2\alpha$

$= x^2 - 2x\cos a + 1 =$ R.H.S.

Example 32:

Solve the equation $z^8 + 1 = 0$ and deduce that

$$\cos 4\theta = 8 \prod_{p=0}^{3}\left[\cos\theta - \cos(2p+1)\frac{\pi}{8}\right]$$

Solution:

$$z^8 + 1 = 0 \Rightarrow z = (-1)^{1/8} = (\cos\pi + i\sin\pi)^{1/8}$$

$$\Rightarrow z = \cos(2p+1)\frac{\pi}{8} + i\sin(2p+1)\frac{\pi}{8},\ p = 0, 1\ldots, 7.$$

$$\Rightarrow z = \cos(2p+1)\frac{\pi}{8} \pm i\sin(2p+1)\frac{\pi}{8},\ p = 0, 1, 2, 3.$$

$$\therefore z^8 + 1 = \prod_{p=0}^{3}\left[z - \left(\cos(2p+1)\frac{\pi}{8} - i\sin(2p+1)\frac{\pi}{8}\right)\right]$$

$$\times\left[z - \left(\cos(2\pi+1)\frac{\pi}{8} - i\sin(2p+1)\frac{\pi}{8}\right)\right].$$

By the above formula, we see that

$$z^8 + 1 = \prod_{p=0}^{3}\left\{z^2 - 2z\cos(2p+1)\frac{\pi}{8} + 1\right\}$$

Dividing both sides of (1) by z^4, we get

$$z^4 + z^{-4} = \prod_{p=0}^{3}\left\{z - 2\cos(2p+1)\frac{\pi}{8} + z^{-1}\right\}. \qquad \ldots(2)$$

On putting $z = \cos\theta + i\sin\theta$ in (2), we get

$$2\cos 4\theta = \prod_{p=0}^{3}\left[2\cos\theta - 2\cos(2p+1)\frac{\pi}{8}\right] \quad \left(\begin{array}{l}\because\ z + z^{-1} = 2\cos\theta \\ z^4 + z^{-4} = 2\cos 4\theta\end{array}\right)$$

$$\text{Hence } \cos 4\theta = 8\prod_{p=0}^{3}\left[\cos\theta - \cos(2p+1)\frac{\pi}{8}\right].$$

Example 33:

Solve the equation $z^{14} - 1 = 0$ and show that:

$$z^{14} - 1 = (z^2 - 1)\prod_{p=1}^{3}\left(z^2 - 2z\cos\frac{p\pi}{7} + 1\right).$$

Hence show that: $\sin 7\theta = 64\sin\theta \prod_{p=1}^{3}\left(\sin^2\frac{p\pi}{7} - \sin^2\theta\right)$

Solution:

$z^{14} - 1 \Rightarrow z = (1)^{1/14} = (\cos 0 + i \sin 0)^{1/14}$

$$\Rightarrow z = \cos \frac{2k\pi}{14} + i \sin \frac{2k\pi}{14};\ k = 0, 1, ...13$$

$$\therefore z = \pm 1, \cos \frac{p\pi}{7} \pm i \sin \frac{p\pi}{7},\ p = 1, 2,..., 6.$$

Now $$z^{14} \Rightarrow 1 = (z^2 - 1) \prod_{p=1}^{6} \left(z^2 - 2z \cos p\pi/7 + 1\right). \quad ...(1)$$

Dividing both sides of (1) by z^7, we get

$$(z^7 - z^{-7}) = (z - z^{-1}) \prod_{p=1}^{6} \left(z - 2 \cos p\pi/7 + z^{-1}\right). \quad ...(2)$$

On putting $z = \cos\theta + i \sin\theta$ in (2), we get

$$2i \sin 7\theta = 2i \sin\theta \prod_{p=1}^{6} \left(2\cos\theta - 2\cos p\pi/7\right) \quad \left(\because \begin{array}{l} z + z^{-1} = 2\cos\theta \\ z - z^{-1} = 2i \sin\theta \end{array}\right)$$

$$\Rightarrow \sin 7\theta = 2^6 \sin\theta \prod_{p=1}^{3} (\cos\theta - \cos p\,\pi/7)(\cos\theta + \cos p\,\pi/7)$$

$$= 2^6 \sin\theta \prod_{p=1}^{3} (\cos^2\theta - \cos^2 p\pi/7)$$

$$= 2^6 \sin\theta \prod_{p=1}^{3} [(1 - \sin^2\theta) - (1 - \sin^2 p\pi/7)]$$

Hence $$\sin 7\theta = 64 \sin\theta \prod_{p=1}^{3} (\sin^2 p\pi/7 = \sin^2\theta).$$

Example 34:

Solve the equation $z^{10} - 1 = 0$ *and deduce that* $\sin 5\theta = 5 \sin\theta$

$$\left[1 - \frac{\sin^2\theta}{\sin^2(\pi/5)}\right]\left[1 - \frac{\sin^2\theta}{\sin^2(2\pi/5)}\right].$$

Solution:

$z^{10} - 1 = 0 \Rightarrow z = (\cos 0 + i \sin 0)^{1/10}$

$$\therefore z = \cos \frac{2p\pi}{10} + i \sin \frac{2\pi p}{10};\ i = 0, 1,....,9$$

$$\Rightarrow z = \cos\frac{p\pi}{5} + i\sin\frac{p\pi}{5};\ i = 0, 1, ..., 9$$

$= \pm 1, \cos p\pi/5 + i \sin p\pi/5;\ p = 1, 2, 3, 4, 5, 6, 7, 8, 9$

$= \pm 1, \cos p\,\pi/5 \pm i \sin p\pi/5;\ p = 1, 2, 3, 4.$

Now $z^{10} - 1 = (z^2 - 1)\prod_{p=1}^{4}(z^2 - 2z\cos p\pi/5 + 1).$ ...(1)

Dividing both sides of (2) by z^5, we get

$(z^5 - z^{-5}) = (z - z^{-1})\prod_{p=1}^{4}(z - 2\cos p\,\pi/5 + z^{-1}).$ ...(2)

Putting $z = \cos\theta + i\sin\theta$ in (2), we get

$$2i\sin 5\theta = 2i\sin\theta\prod_{p=1}^{4}(2\cos\theta - 2\cos p\pi/5)$$

$$\Rightarrow \sin 5\theta = 2^4\sin\theta\prod_{p=1}^{2}(\cos\theta - \cos p\pi/5)(\cos\theta + \cos p\pi/5)$$

$$= 2^4\sin\theta\prod_{p=1}^{2}(\cos^2\theta - \cos^2 p\,\pi/5)$$

$$= 2^4\sin\theta\prod_{p=1}^{2}(\sin^2 p\pi/5 - \sin^2\theta)$$

$$= 2^4\sin\theta(\sin^2\pi/5 - \sin^2\theta)(\sin^2 2\pi/5 - \sin^2\theta)$$

$$= 2^4\sin^2\pi/5.\sin^2 2\pi/5.\sin\theta\left[1 - \frac{\sin^2\theta}{\sin^2(\pi/5)}\right]\left[1 - \frac{\sin^2\theta}{\sin^2(2\pi/5)}\right]$$

$$= 5\sin\theta\left[1 - \frac{\sin^2\theta}{\sin^2(\pi/5)}\right]\left[1 - \frac{\sin^2\theta}{\sin^2(2\pi/5)}\right]$$

$$\left[\because \sin\frac{\pi}{5} = \frac{\sqrt{10 - 2\sqrt{5}}}{4},\ \sin\frac{2\pi}{5} = \frac{\sqrt{10 + 2\sqrt{5}}}{4}\right]$$

Example 35:

Prov that cos (2π/7), cos (4π/7), cos (6π/7) are the roots of the equation $8x^3 + 4x^2 - 4x - 1 = 0.$

Solution:

$\Rightarrow z^7 = 1 \Rightarrow z = (1)^{1/7} = (\cos 0 + i \sin 0)^{1/7}$

$$z = \cos \frac{2k\pi}{7} + i \sin \frac{2k\pi}{7} k = 0, 1, 2, ..., 6$$

$$\therefore z = 1, \cos \frac{2k\pi}{7} \pm i \sin \frac{2k\pi}{7}; k = 1, 2, 3.$$

Using the formula given earlier, wre have

$$z^7 - 1 = (z - 1)[z^2 - 2z \cos (2\pi/7) + 1][z^2 - 2z \cos (4\pi/7) + 1] \times [z^2 - 2z \cos (6\pi/7) + 1].$$

Dividing both sides by $z - 1$, we obtain

$$z^6 + z^5 + z^4 + z^3 + z^2 + z + 1 = (z^2 - 2z \cos 2\pi/7 + 1) \times (z^2 - 2z \cos 4\pi/7 + 1)(z^2 - 2z \cos 6\pi/7 + 1).$$

Dividing the L.H.S. by z^3 and each of the three factors on the R.H.S. by z, we get

$$(z^3 + z^{-3}) + (z^2 + z^{-2}) + (z + z^{-1}) + 1$$
$$= (z + z^{-1} - 2 \cos 2\pi/7)(z + z^{-1} - 2 \cos 4\pi/7)(z + z^{-1} - 2 \cos 6\pi/7). \quad ...(1)$$

Let $z + z^{-1} = 2x$. Then $4x^2 = (z + z^{-1})^2 = z^2 + z^{-2} + 2$

$\Rightarrow z^2 + z^{-2} = 4x^2 - 2.$

Now $8x^3 = (z + z^{-1})^3 = z^3 + z^{-3} + 3zz^{-1}(z + z^{-1}) = z^3 + z^{-3} + 6x$

$\Rightarrow z^3 + z^{-3} = 8x^3 - 6x$. Then equation (1) now becomes

$$(8x^3 - 6x) + (4x^2 - 2) + 2x + 1$$
$$= (2x - 2 \cos 2\pi/7)(2x - 2\cos 4\pi/7)(2x - 2\cos 6\pi/7)$$

$$\Rightarrow 8x^3 + 4x^2 - 4x - 1 = 8(x - \cos 2\pi/7)(x - \cos 4\pi/7)(x - \cos 6\pi/7). \quad ...(2)$$

It follows that $\cos 2\pi/7$, $\cos 4\pi/7$, $\cos 6\pi/7$ are the roots of the equation

$$8x^3 + 4x^2 - 4x - 1 = 0.$$

Example 36:

Solve the equation $x^7 - 1 = 0$, and deduce that

(i) $\cos \frac{2\pi}{7} + \cos \frac{4\pi}{7} + \cos \frac{8\pi}{7} = -\frac{1}{2}$,

(ii) $\cos\frac{2\pi}{7}\cos\frac{4\pi}{7}\cos\frac{8\pi}{7}=\frac{1}{8}$.

Solution:

From equation (2) and keeping in mind the relations between the roots coefficients, it follows that

$$\cos\frac{2\pi}{7}+\cos\frac{4\pi}{7}+\cos\frac{6\pi}{7}=-\frac{4}{8}=-\frac{1}{2},$$

$$\cos\frac{2\pi}{7}\cos\frac{4\pi}{7}\cos\frac{6\pi}{7}=\frac{1}{8}.$$

Example 37:

Solve the equation $z^7+1=0$, and deduce that

$$\cos\frac{\pi}{7}\cos\frac{3\pi}{7}\cos\frac{5\pi}{7}=-\frac{1}{8}.$$

Solution:

$z^7+1=0\Rightarrow z=(-1)^{1/7}=(\cos\pi+i\sin\pi)^{1/7}$

$\Rightarrow z=\cos(2k+1)\frac{\pi}{7}+i\sin(2k+1)\frac{\pi}{7}$; $k=0,1,2,\ldots,6$

$\therefore z=-1, \cos\frac{p\pi}{7}\pm i\sin\frac{p\pi}{7}$, $p=1,3,5$

(For $k=3$, we obtain $z=-1$)

We can write

$$z^7+1=(z+1)(z^2-2z\cos\pi/7+1)(z^2-2z\cos 3\pi/7+1)\times(z^2-2z\cos 5\pi/7+1). \quad ...(1)$$

Dividing both sides by $z+1$, we obtain

$$z^6-z^5+z^4-z^3+z^2-z+1=(z^2-2z\cos\pi/7+1)(z^2-2z\cos 3\pi/7+1)\times(z^2-2z\cos 5\pi/7+1).$$

Dividing the L.H.S. by z^3 and each of the three factors on the R.H.S. by z, we get

$$(z^3+z^{-3})-(z^2+z^{-2})+(z+z^{-1})-1$$

$$=(z+z^{-1}-2\cos\pi/7)(z+z^{-1}-2\cos 3\pi/7)(z+z^{-1}-2\cos 5\pi/7).$$

Putting $z+z^{-1}=2x$, we obtain

$$(8x^3-6x)-(4x^2-2)+2x-1$$

$= 8\ (x - \cos \pi/7)\ (x - \cos 3\pi/7)\ (x - \cos 5\ \pi/7)$

Thus $\cos \pi/7$, $\cos 3\pi/7$, $\cos 5\pi/7$ are the roots of the equation

$$8x^3 - 4x^2 - 4x + 1 = 0.$$

Further $\cos\frac{\pi}{7} + \cos\frac{3\pi}{7} + \cos\frac{5\pi}{7} = \frac{4}{8} = \frac{1}{2}$,

and $\cos\frac{\pi}{7}\cos\frac{3\pi}{7}\cos\frac{5\pi}{7} = -\frac{1}{8}$.

Example 38:

Solve the equation $z^9 + 1 = 0$ and hence obtain the equation whose roots are

$$\cos\frac{\pi}{9}, \cos\frac{3\pi}{9}, \cos\frac{5\pi}{7}, \cos\frac{7\pi}{9}.$$

Solution:

$z^9 + 1 = 0 \Rightarrow z = (\cos \pi + i \sin \pi)^{1/9}$

$\Rightarrow z = \cos\frac{(2k+1)\pi}{9} + i \sin\frac{(2k+1)\pi}{9}$; $k = 0, 1,\ldots, 8$

$\Rightarrow z = -1, \cos\frac{p\pi}{9} \pm i \sin\frac{p\pi}{9}$; $p = 1, 3, 5, 7.$

$z^9 + 1 = (z + 1)\ (z^2 - 2z \cos \pi/9 + 1)\ (z^2 - 2z \cos 3\pi/9 + 1)$

$\times\ (z^2 - 2z \cos 5\ \pi/9 + 1)\ (z^2 - 2z \cos 7\pi/9 + 1)$

Dividing both sides by $z + 1$, we get

$z^8 - z^7 + z^6 - z^5 + z^4 - z^3 + z^2 - z + 1$

$= (z^2 - 2z \cos \pi/9 + 1)\ (z^2 - 2z \cos 3\pi/9 + 1)$

$\times\ (z^2 - 2z \cos 5\ \pi/9 + 1)\ (z^2 - 2z \cos 7\pi/9 + 1).$

Dividing the L.H.S. by z^4 and each of the four factors on the R.H.S. by z, we get

$(z^4 + z^{-4}) - (z^3 + z^{-3}) + (z^2 + z^{-2}) - (z + z^{-1}) + 1$

$= (z + z^{-1} - 2 \cos \pi/9)\ (z + z^{-1} - 2 \cos 3\pi/9)$

$(z + z^{-1} - 2 \cos 5\pi/9) \times (z + z^{-1} - 2 \cos 7\pi/9).$

Putting $z + z^{-1} = 2x$, we get

$[16x^4 - 4(4x^2 - 2) - 6] - (8x^3 - 6x) + (4x^2 - 2) - 2x + 1$

$= 2^4\ (x - \cos \pi/9)\ (x - \cos 3\pi/9)\ (x - \cos 5\pi/9)\ 9x - \cos 7\pi/9).$

Hence $\cos \pi/9$, $\cos 3\pi/9$, $\cos 5\pi/9$, $\cos 7\pi/9$ are the roots of the equation

$$16x^4 - 16x^2 + 2 - 8x^3 + 6x + 4x^2 - 2x - 1 = 0$$

$$\Rightarrow \quad 16x^4 - 8x^3 - 12x^2 + 4x + 1 = 0.$$

Example 39:

Solve the equation $z^7 + 1 = 0$, and show that

$$(1 + y)^7 + (1 - y)^7 = 14\left(y^2 + \tan^2\frac{\pi}{14}\right)\left(y^2 + \tan^2\frac{3\pi}{14}\right)\left(y^2 + \tan^2\frac{5\pi}{14}\right)$$

Deduce that: $\tan^2\frac{\pi}{14} + \tan^2\frac{3\pi}{14} + \tan^2\frac{5\pi}{14} = 5$.

Solution:

The equation (2) of Example can be written as

$$z^7 + 1 = (z + 1)\prod_{p=0}^{2}\left[z^2 - 2z\cos(2p+1)\frac{\pi}{7} + 1\right]$$

Putting $z = \frac{1+y}{1-y} \Rightarrow z^7 + 1 = \frac{(1+y)^7 + (1-y)^7}{(1-y)^7}$,

$$z^2 + 1 = \frac{(1+y)^2 + (1-y)^2}{(1-y)^2} = \frac{2(1+y^2)}{(1-y)^2},\ z + 1 = \frac{2}{1-y}.$$

Using these relations in (2), we obtain

$$\frac{(1+y)^7 + (1-y)^7}{(1-y)^7} = \left(\frac{2}{1-y}\right)\prod_{p=0}^{2}\left[\frac{2(1+y^2)}{(1-y)^2} - \frac{2(1+y)}{(1-y)}\cos(2p+1)\frac{\pi}{7}\right]$$

$$= \left(\frac{2}{1-y}\right)\left[\frac{2}{(1-y)^2}\right]^3\prod_{p=0}^{2}\left[(1+y^2) - (1-y^2)\cos(2p+1)\frac{\pi}{7}\right]$$

$\Rightarrow (1 + y)^7 + (1 - y)^7$

$$= 2^4\prod_{p=0}^{2}\left[\left\{1 - \cos(2p+1)\frac{\pi}{7}\right\} + y^2\left\{1 + \cos(2p+1)\frac{\pi}{7}\right\}\right]$$

$$= 2^4\prod_{p=0}^{2}\left[2\sin^2(2p+1)\frac{\pi}{14} + 2y^2\cos^2(2p+1)\frac{\pi}{14}\right]$$

$$= 2^7\prod_{p=0}^{2}\left[\sin^2(2p+1)\frac{\pi}{14} + y^2\cos^2(2p+1)\frac{\pi}{14}\right]. \quad ...(3)$$

Dividing both sides of (2) by z + 1, we obtain

$$z^6 - z^5 + z^4 - z^3 + z^2 - z + 1 = \prod_{p=0}^{2}\left[z^2 - 2z\cos(2p+1)\frac{\pi}{7} + 1\right]$$

Putting z = – 1 in the above equation, we obtain

$$7 = \prod_{p=0}^{2} 2\left\{1+\cos(2p+1)\frac{\pi}{7}\right\} = \prod_{p=0}^{2} 2^2\cos^2(2p+1)\frac{\pi}{14}$$

$$\Rightarrow 7 = 2^6 \prod_{p=0}^{2}\cos^2(2p+1)\frac{\pi}{14}. \qquad ...(4)$$

Dividing both sides of (3) by those of (4), we get

$$(1+y)^7 + (1-y)^7 = 14\prod_{p=0}^{2}\left[y^2 + \tan^2(2p+1)\frac{\pi}{14}\right]$$

$$= 14\left(y^2+\tan^2\frac{\pi}{14}\right)\left(y^2+\tan^2\frac{3\pi}{14}\right)\left(y^2+\tan^2\frac{5\pi}{14}\right).$$

Equating the coefficients of y^4 on both sides, we have $2\left(7_{c4}\right) = 14$ $\left(\tan^2\frac{\pi}{14} + \tan^2\frac{3\pi}{14} + \tan^2\frac{5\pi}{14}\right)$

Hence $\tan^2\frac{\pi}{14} + \tan^2\frac{3\pi}{14} + \tan^2\frac{5\pi}{14} = \frac{1}{7}\left(\frac{7!}{4!\,3!}\right) = 5.$

Example 40:

Solve the equation $(1+w)^9 - (1-w)^9 = 0$ and hence obtain the equation whose roots are

$$\tan^2\frac{\pi}{9},\ \tan^2\frac{2\pi}{9},\ \tan^2\frac{3\pi}{9},\ \tan^2\frac{4\pi}{9}$$

Deduce that: $\tan\frac{\pi}{9}\tan\frac{2\pi}{9}\tan\frac{3\pi}{9}\tan\frac{4\pi}{9} = 3.$

Solution:

Consider $z^9 - 1 = 0 \Rightarrow z = (\cos 0 + i\sin 0)^{1/9}$

$\Rightarrow z = \cos 2p\pi/9 + i\sin 2p\pi/9,\ p = 0, 1, \ldots, 8$

$\Rightarrow z = 1,\ \cos 2p\pi/9 \pm i\sin 2p\pi/9,\ p = 1, 2, 3, 4.$

$\therefore\ z^9 - 1 = (z - 1) \prod_{p=1}^{4} (z^2 - 2z \cos 2p\pi/9 + 1).$

Putting $z = \dfrac{1+w}{1-w}$, we obtain

$$\left(\frac{1+w}{1-w}\right)^9 - 1 = \left(\frac{1+w}{1-w} - 1\right) \prod_{p=1}^{4} \left[\left(\frac{1+w}{1-w}\right)^2 - 2\left(\frac{1+w}{1-w}\right) \cos \frac{2p\pi}{9} + 1\right]$$

$$\Rightarrow (1 + w)^9 - (1 - w)^9 = 2^5\ w \prod_{p=1}^{4} \left[(1 + w^2) - (1 - w^2) \cos \frac{2p\pi}{9}\right]$$

$$= 2^5\ w \prod_{p=1}^{4} \left[\left(1 - \cos \frac{2p\pi}{9}\right) + w^2 \left(1 + \cos \frac{2p\pi}{9}\right)\right]$$

$$\therefore\ (1 + w)^9 - (1 - w)^9 = 2^9\ w \prod_{p=1}^{4} \left[\sin^2 \frac{p\pi}{9} + w^2 \cos^2 \frac{p\pi}{9}\right]. \quad ...(1)$$

Now $(1 + w)^9 - (1 - w)^9$

$$\left(1 + 9_{c1} w + 9_{c2} w^2 + 9_{c3} w^3 + 9_{c4} w^4 + 9_{c5} w^5 + 9_{c6} w^6 + 9_{c7} w^7 + 9_{c8} w^8 + w^9\right)$$

$$-\left(1 - 9_{c1} w + 9_{c2} w^2 - 9_{c3} w^3 + 9_{c4} w^4 - 9_{c5} w^5 + 9_{c6} w^6 - 9_{c7} w^7 + 9_{c8} w^8 - w^9\right)$$

$$= 2\ \left(9_{c1} w + 9_{c3} w^3 + 9_{c5} w^5 + 9_{c7} w^7 + w^9\right)$$

$$= 2\ (9w + 84w^3 + 126w^5 + 36w^7 + w^9). \quad ...(2)$$

Putting (2) in (1) and dividing by 2w, we get

$$9 + 84w^2 + 126w^4 + 36w^6 + w^8 = 2^8 \prod_{p=1}^{4} \left[\sin^2 \frac{p\pi}{9} + w^2 \cos^2 \frac{p\pi}{9}\right]$$

$$= 2^8 \prod_{p=1}^{4} \cos^2 \frac{p\pi}{9} \left[w^2 + \tan^2 \frac{p\pi}{9}\right] \quad ...(3)$$

Putting $w^2 = -z$ in (3); we see that

$\tan^2 (\pi/9)$, $\tan^2 (2\pi/9)$, $\tan^2 (3\pi/9)$, $\tan^2 (4\pi/9)$ are the roots of

$$9 - 84z + 126z^2 - 36z^3 + z^4 = 0$$

$$\Rightarrow z^4 - 36z^3 + 126z^2 - 84z + 9 = 0. \quad ...(4)$$

(ii) From equation (4), it follows that

$$\tan^2 \frac{\pi}{9} \tan^2 \frac{2\pi}{9} \tan^2 \frac{3\pi}{9} \tan^2 \frac{4\pi}{9} = 9.$$

Hence $\tan\frac{\pi}{9}\tan\frac{2\pi}{9}\tan\frac{3\pi}{9}\tan\frac{4\pi}{9}=3$.

Example 41:

Solve the equation $z^9 - 1 = 0$, and deduce that

(i) $\sin\frac{\pi}{9}\sin\frac{2\pi}{9}\sin\frac{3\pi}{9}\sin\frac{4\pi}{9}=\frac{3}{16}$,

(ii) $\tan\frac{\pi}{9}\tan\frac{2\pi}{9}\tan\frac{3\pi}{9}\tan\frac{4\pi}{9}=3$.

Solution:

We have seen that

$$\tan\frac{\pi}{9}\tan\frac{2\pi}{9}\tan\frac{3\pi}{9}\tan\frac{4\pi}{9}=3 \qquad ...(5)$$

Comparing the coefficients w^8 on both sides of (3), we get

$$1 = 2^8\cos^2\frac{\pi}{9}\cos^2\frac{2\pi}{9}\cos^2\frac{3\pi}{9}\cos^2\frac{4\pi}{9}$$

$$\Rightarrow \cos\frac{\pi}{9}\cos\frac{2\pi}{9}\cos\frac{3\pi}{9}\cos\frac{4\pi}{9}=\frac{1}{2^4}=\frac{1}{16}. \qquad ...(6)$$

Multiplying the respective sides of (5) and (6), we get

$$\sin\frac{\pi}{9}\sin\frac{2\pi}{9}\sin\frac{3\pi}{9}\sin\frac{4\pi}{9}=\frac{3}{16}.$$

Example 42:

Solve the equation $(1+z)^{11}-(1-z)^{11}=0$, and hence obtain the equation whose roots are

$$\tan^2\frac{\pi}{11}, \tan^2\frac{2\pi}{11},\ldots, \tan^2\frac{5\pi}{11}.$$

Solution:

Consider $x^{11} - 1 = 0$

$$\Rightarrow x = \cos\frac{2p\pi}{11}+i\sin\frac{p\pi}{11}, \; p = 0, 1,\ldots, 10$$

$$\Rightarrow x = 1, \cos\frac{2p\pi}{11}\pm i\sin\frac{2p\pi}{11}; \; p = 1, 2, 3, 4, 5.$$

$$\therefore x^{11} - 1 = (x - 1) \prod_{p=1}^{5} \left(x^2 - 2x \cos \frac{2p\pi}{11} + 1\right).$$

Putting $x = \dfrac{1+z}{1-z}$ we obtain

$$\left(\frac{1+z}{1-z}\right)^{11} - 1 = \left(\frac{1+z}{1-z} - 1\right) \prod_{p=1}^{5} \left\{\left(\frac{1+z}{1-z}\right)^2 - 2\left(\frac{1+z}{1-z}\right) \cos \frac{2p\pi}{11} + 1\right\}$$

$$\Rightarrow (1 + z)^{11} - (1 - z)^{11} = 2^6.z \prod_{p=1}^{5} \left\{(1+z^2) - (1-z^2) \cos \frac{2p\pi}{11}\right\}$$

$$\Rightarrow (1 + z)^{11} - (1 - z)^{11} = 2^{11}.z \prod_{p=1}^{5} \left\{\sin^2 \frac{p\pi}{11} + z^2 \cos^2 \frac{p\pi}{11}\right\}.$$

Expanding $(1 + z)^{11}, (1 - z)^{11}$ by Binomial theorem and dividing throughout by 2z, we obtain

$$11 + 165z^2 + 462z^4 + 330z^6 + 55z^8 + z^{10}$$

$$= 2^{10} \prod_{p=1}^{5} \left\{\sin^2 \frac{p\pi}{11} + z^2 \cos^2 \frac{p\pi}{11}\right\}$$

$$= 2^{10} \prod_{p=1}^{5} \cos^2 \frac{p\pi}{11} \left\{\tan^2 \frac{p\pi}{11} + z^2\right\}.$$

Put $z^2 = -x$, we see that

$\tan^2 \dfrac{\pi}{11}, \tan^2 \dfrac{2\pi}{11}, \ldots, \tan^2 \dfrac{5\pi}{11}$ are the roots of the equation

$$11 - 165x + 462x^2 - 330x^3 + 55x^4 - x^5 = 0$$

$$\Rightarrow x^5 - 55x^4 + 330x^3 - 462x^2 + 165x - 11 = 0.$$

EXERCISES

1. Solve $x^6 + 1 = 0$.
2. Solve $x^7 - 1 = 0$.
3. Prove that the sum of the pth powers of the roots of the equation $x^n = 1$, where n is a positive integer is 0 when p is not a multiple of n and n when p is a multiple of n.
4. Find the seven seventh roots of unity and prove that these roots are in geometrical progression and the sum of their powers always vanishes unless n is a multiple of 7 (n being any integer).

5. Prove that, if x denotes any nth roots of unity, then $1 + x + x^2 + x^3 + \ldots + x^n = 0$.

6. Show that

   $\sin 6\theta = 6 \cos^5 \theta \sin \theta - 20 \cos^3 \theta \sin^3 \theta + 6 \cos \theta \sin^5 \theta$.

7. Express $\cos 8\theta$ in terms of powers of $\cos \theta$.

8. Show that $(\sin n\theta)/\sin \theta$

   $$= (2 \cos \theta)^{n-1} - (n-2)(2 \cos \theta)^{n-3} + \frac{(n-3)(n-4)}{1.2}(2 \cos \theta)^{n-5} + \ldots + \text{last term}$$

   Giving reasons show that the last term is $(-1)^{(n-1)/2}$ if n is odd and $(-1)^{(n/2)-1}(n \cos \theta)$ if n is even

9. Show that $\sin \theta \cos^2 \theta = \dfrac{1}{4}\left[(1+3)\theta - (1+3^3)\dfrac{\theta^3}{3!} + (1+3^5)\dfrac{\theta^5}{5!} - \ldots\right]$.

10. Prove that $\cos^3 x = \dfrac{1}{4}\left[x - \dfrac{x^2}{2!}(3^2+3) + \dfrac{x^4}{4!}(3^4+3) - \ldots\right]$.

11. Show that

    $$\sin^3\theta = \frac{3}{4}\left[\frac{(3^2-1)}{3!}\theta^3 - \frac{(3^4-1)}{5!}\theta^5 + \ldots + \frac{(-1)^{n-4}(3^{2n}-1)}{(2n+1)!}\theta^{2n-1} + \ldots\right]$$

12. If $\dfrac{\sin \theta}{\theta} = \dfrac{863}{864}$, show that $\theta = 4°47'$ nearly.

13. If $\dfrac{\sin \theta}{\theta} = \dfrac{1349}{1350}$, prove that the angle q is very nearly equal to $\dfrac{1}{15}$ radians.

14. Prove that:

    $$\cos\frac{\pi}{11} + \cos\frac{3\pi}{11} + \cos\frac{5\pi}{11} + \cos\frac{7\pi}{11} + \cos\frac{9\pi}{11}, \cos\frac{11\pi}{13} = \frac{1}{2}$$

15. From the equation whose roots are

    $$\cos\frac{\pi}{13}, \cos\frac{3\pi}{13}, \cos\frac{5\pi}{13}, \cos\frac{7\pi}{13}, \cos\frac{9\pi}{13}, \cos\frac{11\pi}{13}$$

16. Prove that $64 \sin^5 \theta \cos^2 \theta = \sin 7\theta - 3 \sin 5\theta + \sin 3\theta + 5 \sin \theta$

17. Show that

    $64 \cos^7 \theta = \cos 7\theta + 7 \cos 5\theta + 21 \cos 3\theta + 35 \cos \theta$.

18. Prove that

$$\sin^6 \theta = -\frac{1}{32} [\cos 6\theta - 6 \cos 4\theta + 15 \cos 2\theta - 10].$$

19. Expand $\sin^5 \theta \cos^2 \theta$ in a series of sines of multiples of θ.
20. Express $\cos^3 \theta \sin^6 \theta$ as a series of cosines of multiples of θ.
21. Expand $\sin^3 \theta \cos^3 \theta$ in a series of sines of multiples of θ.
22. Expand $\cos^6 \theta \sin^3 \theta$ in a series of cosines of multipls of θ.
23. Expand $\sin^3 \theta \cos^8 \theta$ in a series of sines of muliples of θ.
24. Prove that $\cos^6 \theta \sin^2$

$$= -\frac{1}{128} [\cos 8\theta + 4 \cos 6\theta + 4 \cos 4\theta - 4 \cos 2\theta - 5].$$

25. Express $\dfrac{(\cos \theta + i \sin \theta)^7}{(\cos 2\theta + i \sin 2\theta)^2}$ in the form x + iy.

[**Ans.** $\cos 3\theta + i \sin 3\theta$]

26. Express $\dfrac{\{\cos \pi/4 + i \sin \pi/4\}^5}{\{\cos \pi/3 + i \sin \pi/3\}^4}$ in the form x + iy.

[**Ans.** $\cos (\pi/12) - i \sin (\pi/12)$]

27. Simplify $\dfrac{(\cos 2\theta + i \sin 2\theta)^3 (\cos \theta - i \sin \theta)^{-4}}{(\cos 3\theta + i \sin 3\theta)^4 (\cos 4\theta + i \sin 4\theta)^{-2}}$

[**Ans.** $\cos 6\theta + i \sin 6\theta$]

28. If $2 \cos \theta = \alpha + 1/a$, $2 \cos \phi = \beta + 1/b$; show that

$$\frac{1}{2}\left(ab + \frac{1}{ab}\right) = \cos (\theta + \phi).$$

29. Prove tha $\{(\cos \alpha + i \sin \alpha) - (\cos \beta + i \sin \beta)\}^8 +$

$$\{(\cos \alpha - i \sin \alpha) - (\cos \beta - i \sin \beta)\}^8 = 2^8 \frac{1}{2} (\alpha - \beta) \cos 4 (\alpha + \beta).$$

30. Prove Summation of Series when index is a negative integer.
31. $z^6 = 1$. Ans. $\pm 1, \pm \frac{1}{2} \pm i \left(\sqrt{3/2}\right)$.
32. $z^7 + 1 = 0$. [**Ans.** $\cos (p\pi/7) + i \sin (p\pi/7)$, p=1, 3, 5, 7, 9, 11, 13]
33. $z^8 + 1 = 0$. [**Ans.** $\cos(p\pi/8) + i \sin(p\pi/8)$, p=1, 3, 5, 7, 9, 11, 13, 15]
44. $z^4 + z^3 + z^2 + 1 = 0$.

[**Hint:** L.H.S. = $(z^5 - 1)/(z - 1)$. Now solve $z^5 - 1 = 0$ and discard the value z = 1].

Ans. $\cos(p\pi/5) + i \sin(p\pi/5)$, p = 2, 4, 6, 8.

35. $x^8 - x^6 + x^4 - x^2 + 1 = 0$.

[**Hint:** Multiply by $(x^2 + 1)$ to get $x^{10} + 1 = 0$ and reject the values given by $x^2 + 1 = 0$ *i.e.*, $\pm$ i]

Ans. $\cos(p\pi/10) + i \sin(p\pi/10)$; p = 1, 3, 7, 9, 11, 13, 17, 19.

36. $x^7 + x^4 + x^3 + 1 = 0$.

[**Hint:** L.H.S. = $(x^4 + 1)(x^3 + 1)$.

Ans. $-1, \frac{1}{2} \pm i\left(\sqrt{3/2}\right); \frac{1}{\sqrt{2}}(\pm i \pm i)$.

37. $(z + 1)^5 + (z - 1)^5 = 0$. **Ans.** 0, $\pm i \cot(p\theta/10)$; p = 1, 3.

38. $(4 + z)^5 - (4 - z)^5 = 0$ **Ans.** 0, $\pm 4i \cot(\pi/10)$, $\pm 4i \cot(3\pi/10)$.

39. If $x = \cos(2\pi/5) + i \sin(2\pi/5)$, show that $1 + x^n + x^{2n} + x^{3n} + x^{in}$ is 5 when n is multiple of 5 and 0 when n is not a multiple of 5.

40. Solve $x^m = 1$, m being a positive integer, and show that the roots are in G.P. Hence, show that the sum of qth powers of the roots is m or zero according as θ is a multiple of m or not.

41. $\cos 4\theta = \cos^4\theta - 6\cos^2\theta \sin^2\theta + \sin^4\theta$.

42. $\sin 4\theta = 4\cos^3\theta \sin\theta - 4\cos\theta \sin^3\theta$.

43. $\tan 4\theta = \dfrac{4\tan\theta - 4\tan^3\theta}{1 - 6\tan^2\theta + \tan^4\theta}$.

44. $8\sin^4\theta = \cos 4\theta - 4\cos 2\theta + 3$.

45. $8\cos^4\theta = \cos 4\theta + 4\cos 2\theta + 3$.

46. $64\sin^7\theta = -\sin 7\theta + 7\sin 5\theta - 21\sin 3\theta + 35\sin\theta$.

47. $\cos 2\theta + \cos 4\theta + \cos 6\theta + ...$ **Ans.** $\sin n\theta \cos(n + 1)\theta \operatorname{cosec}\theta$

48. $\sin 2\theta + \sin 4\theta + \sin 6\theta + ...$ **Ans.** $\sin n\theta \sin(n + 1)\theta \operatorname{cosec}\theta$.

49. $\sin(\pi/2n) + \sin(3\pi/2n) + \sin(5\pi/2n) + ...$ **Ans.** $\operatorname{cosec}(\pi/2n)$

50. $\sin\theta + \frac{1}{3}\sin 2\theta + \frac{1}{3^2}\sin 3\theta + ...$

Ans. $[\sin n\theta - 3\sin(n + 1)\theta + 3^{n+1}\sin\theta] \div 3^{n-1}.2(5 - 3\cos\theta)$.

51. $2\cos 2\theta + 2^2\cos 4\theta + 2^3\cos 6\theta + ...$

[**Hint:** $C + iS = z + z^2 + z^3 + ... + z^n$, where $z = 2e^{2i\theta}$.

Ans. $2[\cos 2\theta - 2 - 2^n\cos 2(n + 1)\theta + 2^{n+1}\cos 2n\theta]$
$\div (5-4\cos 2\theta)$.

52. Sum the series: $\cos\alpha + \frac{1}{2}\cos 2\alpha + \frac{1}{2^2}\cos 3\alpha + \ldots +$ ad. inf.

[**Hint:** $C + iS = e^{ia} + \frac{1}{2}e^{2i\alpha} + \frac{1}{2^2}e^{3i\alpha} + \ldots +$ ad. inf.]

This is a geometric series with c.r. $= \frac{1}{2}e^{i\alpha}$

$$\therefore C + iS = \frac{e^{i\alpha}}{1-\frac{1}{2}e^{i\alpha}} = \frac{2(\cos\alpha + i\sin\alpha)}{2-(\cos\alpha + i\sin\alpha)}$$

$$= \frac{(2\cos\alpha + 2i\sin\alpha)[(2-\cos\alpha)+i\sin\alpha]}{(2-\cos\alpha)^2 + \sin^2\alpha = 5-4\cos\alpha} \quad \ldots(1)$$

$\therefore C = [2\cos\alpha(2-\cos\alpha) - 2\sin^2\alpha]/(5-4\cos\alpha)$

$= (4\cos\alpha - 2)/(5-4\cos\alpha)$.

53. Sum the series: $\sin\alpha + \frac{1}{2}\sin 2a + \frac{1}{2^2}\sin 3\alpha + \ldots$ ad inf.

[**Hint:** Equate imaginary parts on both sides of (1)]

Ans. $4\sin a/(5-4\cos a)$.

54. Sum the series: $\cos\frac{\pi}{3} + \frac{1}{3}\cos\frac{2\pi}{3} + \frac{1}{5}\cos\frac{3\pi}{5} + \ldots +$ ad. inf.

Ans. $\frac{1}{8}\left[2\sqrt{3}\log\left(2+\sqrt{3}\right)-\pi\right]$

55. If $a = \cos 2\alpha + i\sin 2\alpha$, with similar expressions for b and c, prove that $\sqrt{(abc)} + \left[1/\sqrt{(abc)}\right] = 2\cos(\alpha+\beta+\gamma)$.

56. Simplify $\frac{(\cos\theta + i\sin\theta)^{10}}{(\cos\alpha + i\sin\alpha)^{12}}$

Ans. $\cos(10-\theta-12\alpha) + i\sin(10-\theta-12\alpha)$

57. Show that $[(\cos\theta - \cos\phi) + i(\sin\theta - \sin\phi)]^n$

$+ [(\cos\theta + \cos\phi) - i(\sin\theta - \sin\phi)]^n$

$= 2^{n+1}\sin^n\left[\frac{1}{2}(\theta-\phi)\right] \times \cos\frac{1}{2}n(\pi+\theta+\phi)$.

58. Form an equation whose roots are cubes of the roots of the equation $x^2 + 2x\cos\theta + 1 = 0$.

59. If $x + (1/x) = 2\cos\theta$, show that

$x^n + (1/x^n) = 2\cos n\theta$.

60. Summation of Series holds good for:

(i) integral indices only;

(ii) rational indices only;

(iii) real indices only;

(iv) complex indices only;

(v) none of the above statements is true. **Ans.** (ii)

61. Discuss all the values of $(\cos\theta + i\sin\theta)^{1/\theta}$, where θ is positive integer.

62. If A and B are the roots $x^2 + 4 = 2x$, find the value of $A^n + B^n$.

63. Solve the equation $4x^4 = \sqrt{3+i}$.

Ans. $\frac{1}{2^{1/4}}\left[\cos\frac{r\pi}{24} + i\sin\frac{r\pi}{24}\right]$, where r = 1, 13, 25. 37

$\left[\textbf{Hint. } x^4 = \frac{\sqrt{3+i}}{4} = \frac{1}{2}\left(\frac{\sqrt{3}}{2} + i.\frac{1}{2}\right) = \frac{1}{2}\left(\cos\frac{\pi}{6} + i\sin\frac{\pi}{6}\right)\right]$

64. Find all the values of $(-8)^{1/6}$.

65. Find all the values of $\left(\sqrt{3i} + 3i\right)^{1/3}$.

66. Find all the values $\left(1 - i\sqrt{3}\right)^{1/3}$.

67. Find all the values of $\left(1 - i\sqrt{3}\right)^{1/5}$.

68. Find all the values of $(-i)^{1/6}$.

69. Solve $x^5 = 1$ by Summation of Series.